Working Horses:

Tradition, Innovation and Sustainability

William Castle

Published by Field and Lane Press, Shropshire, UK.
working-horses.com

Ordering Information:
Copies of this book can be ordered at working-horses.com
For trade orders contact william@working-horses.com

Illustrations are by the author.

Castle, William.
Working Horses: Tradition, Innovation and Sustainability

First Edition 2025

ISBN 978-1-9192415-0-0

Acknowledgements

This book would not have happened had it not been for Lynn Miller, editor of the Small Farmer's Journal and Diana Zeuner, the editor of the Heavy Horse World magazine. In giving space for my writing, they provided me with a channel for my knowledge, thoughts, and experiences to reach others with an interest in working horses, but I also greatly appreciate their commitment and, as an author, their balanced advice and support. I would also like to thank Erhard Schroll for his similar role in editing Starke Pferde magazine, and to Diana and Erhard, who hold part copyright, for permission to use articles that appeared in their magazines.

I would especially like to thank Lynn Miller for writing the Foreword, Robert Sampson, a farmer with a deep understanding of horses and horse work, for our quarterly conversations, for his guidance, good sense and good humour, and Jonathan and Fiona Waterer for their encouragement and whose work with draught horses plays a significant role in this book and in the real world.

I would also like to place on record my appreciation of two horsemen featured in this book who are no longer with us, Geoff Morton and Joe Godderidge, and to thank Andrew Morton and Mark Morton, particularly for giving me a good start with working horses. Thanks also go to Phil Wilson, Kai-Thomas Roth, the staff of the Small Farmer's Journal and the archivists at the Museum of English Rural Life in Reading, who in various ways have contributed to this book.

Without the photographs, this book would be much less interesting and useful, so I would like to thank those who have provided me with photographs - Guzelian Photography, Bournemouth News and Picture Service, Frankie Woodgate, David Mason, Mike Bingham, Diana Zeuner, Tony Johnson, Erhard Schroll, Stephen Anderson-Jones, Fiona Waterer and Mark Morton. Some photos, such as the pictures given by Joe Godderidge and Geoff Morton, are unfortunately without attribution, but I would like to thank everyone else who took the photographs. (listed on the next page)

Lastly, I want to thank my family for their acceptance of my preoccupations, for helping, particularly at haytime, and for taking many of the photographs in this book. My wife, Liz, particularly deserves an additional and huge thank you for carefully reading through numerous drafts and offering an objective overview of the book as it progressed.

Photographic Credits

Will Anderson-Jones - page 96.

Mike Bingham - page 86, both pictures.

Isabel Castle - page 46.

Liz Castle - pages 41, 43, 44 bottom, 47, 49, 55 both, 56 both, 59, 64 bottom, 65, 66, 94, 95, 101.

Samuel Castle - front cover and pages 3, 70, 78, 87, 155, 160, 168.

William Castle - pages 6, 8, 9, 10, 11 top, 14, 15, 17, 19, 21, 30 bottom, 33, 34, 36, 37, 42, 44 top, 45, 50, 51, 52 both, 57, 59 top and bottom two, 62, 64 top, 67 top, 68 both, 69, 72, 73, 75, 76, 81, 82, 83 both, 89, 92, 96 top, 98 both, 99 both, 100 top, 102, 103, 104 both, 105, 106, 115, 121, 122, 126, 127, 129, 132, all of Chapter Ten, 147, 148, 149, 150, 153, 158, 167.

Zachary Culpin - page 5

Kristi Gilman-Miller - page 1.

Tony Johnson - page 18.

David Mason - pages 97, 151.

Andrew Morton - pages 11 bottom, 12, 13, 93, 107, 113.

Justin Slee - page 91.

Erhard Schroll - page 67 bottom.

Bob Atkins - page 100 bottom.

Fiona Waterer - pages 23, 25, 26 both, 28, 29, 30 top, 31, 32, back cover top.

Diana Zeuner - page 35.

Front Cover photo - The author's pair of Percheron mares.

Back Cover - Jonathan Waterer rowing up hay with a New Holland side-delivery rake behind a hitch cart, and a young Amish woman discing and rolling in Pennsylvania.

Table of Contents

Foreword by Lynn R Miller		Page 1
Preface		3
Introduction		5
Chapter 1 -	A Farm Powered by Horses	7
Chapter 2 -	Two More Farms	23
Chapter 3 -	Making Hay with a Single Horse	41
Chapter 4 -	Comparative Craftsmanship	69
Chapter 5 -	Learning to Plough	75
Chapter 6 -	Good Horses	91
Chapter 7 -	Ploughing with a Single Horse	101
Chapter 8 -	You Can't Work Horses and Send Men to the Moon!	107
Chapter 9 -	The Pull of the Horse-drawn Plough	113
Chapter 10 -	Lancaster County	133
Chapter 11 -	The Rhythm of Horses in the Landscape	147
Chapter 12 -	The Spark	153
References		157
Recommended Reading		159
Glossary		161
About the Author		168

Opposite - Lynn Miller setting out to rake hay with two Belgians and an American pull-from-behind buckrake.

Foreword

A Role for the Working Horse

This book is a remarkable and insightful gift to the future from a consummate craftsman and teamster. **Working Horses,** in powerful and closely held ways, points to a brighter horizon for small farms anywhere.

We are here, deep within a destructive and cancerous period in human history. Our planet, our climate, our communities need calm, certain and determined messengers, workable examples, prophetic voices, artisans, teachers, janitors, coaches and caregivers to show us better ways forward. William Castle - Luthier, Farmer, Horseman, Artist - fits that bill appropriately and with tool-rack and table-top intrigue. This book applies to our general need of good hope and, along the way, we, some of us, may find this volume to be larger than its subject. I do.

Should you be one who has an interest in the immediacy and application of the teamster's craft, this book may well be essential. And if you happen to be someone already invested in learning the ropes, so to speak, William Castle's Working Horses will be as if a godsend, it will make the journey easier and more enjoyable, it will give your work animals a better chance at a pleasant, defining work existence.

And all of that is because this book comes from an ideal wellspring. For thirty plus years Castle has operated his soft step *'holding'* with a dependence upon, and enjoying the benefits of, truest horsepower.

My own relationship with working horses gifted me a character-shaping string of epiphanies and gratitudes. 'I know the feeling' at the end of a long day spent in union with working equines, how in those many, many moments I am thrilled by what has been accomplished and by how it was done. I read of that feeling here in this book - this manual - throughout William's writing.

William Castle is a craftsman. As such he is steeped in the traditions of work process, of the fit and effectiveness of tools, of the grain of each circumstance, of the integration and protection of heritage, of the next earned comfort. All of these things and much more go into the assessment that here is a man who is best suited to step up as guide and instructor of this way of working. In an artisan's manner Castle accumulated, over his long service with work horses, measurements of the appropriate and of the transformative tools and fittings. He presents those here in careful order and, ultimately, in a form that allows intelligent pilgrims access to the wondrous craft of working horses and mules in harness.

Though Castle's purpose is to share the how to's, the clarity and usefulness of this book also serves as an answer to those who might feel compelled to say 'Working Horses aren't viable. Working horses do not belong in the future.' Working horses **are** a viable motive power source. And working horses not only belong in our future, they may help to define it as an improvement.

The words are different, the outward expressions vary, but in 2025 many people spread across the planet's arable lands pine for a landed venture the scale of which suits their temperament, their resources, their goals. 'Scale' defines these dreams, human scale and a pace that matches the heartbeat. Synchronicity of pace multiplies capability in ways that defy normal accounting. Work alongside horses or mules and you will come to feel that your aggregate numbers are less than your combined strength and vigor. You want it. It wants you. The way to get there is with information you might trust and believe in. William's book offers this in spades.

— Lynn R. Miller, Singing Horse Ranch, Oregon

Preface

August 2025

From my very first experience of driving a horse I was intrigued by the process, fascinated by the horses' capabilities, by the interaction between horse and human and the knock-on effects that using horses had on the land.

As a beginner, I had a lot to learn, so besides taking every opportunity to work horses, I was always asking questions, always wanting to know more. It was that inquisitiveness which led me to start writing articles, mostly as an excuse to visit other horse farmers and ask them a whole load of questions! In time, I started writing about my own experiences, to explore the practical constraints and thinking processes of starting out with horses on one's own. Later, my attention turned more towards how horses were used in the past, by researching in archives, trawling through books, magazines and catalogues from the nineteenth and early twentieth centuries, which in turn became the starting point for further articles.

Over the years, those articles have amounted to many thousands of words, this book being a selection of those I believe are sufficiently interesting, relevant or different to have some enduring value. Although some time has passed since the earlier articles first appeared and two of the featured horsemen, Joe Godderidge and Geoff Morton, are no longer with us, within the context of horse work, of farming and our relationship with the wider world, those words are just as pertinent today as when first written.

When we use horses for work, in the beginning we need to know how to do it, but to fully understand what is happening we need to ask why. Sometimes that 'why' may be relatively straightforward - why do you harness a horse in a particular order or arrange the individual horses in a team in a certain manner, but ask a few more questions and the 'whys' soon encompass wider concerns - why do you find mixed cropping on horse-powered farms, why is soil farmed with horses easier to work, and, for me at least, why are we not using many more horses than we do?

The chapters that first appeared as guest editorials in the Small Farmer's Journal take a broad view of these questions, drawing in diverse influences, casting light on farming and working horses in relation to the wider environment, writing which is somewhat philosophical but is always grounded in practicality. Other chapters, by contrast, concentrate on the hands-on experience and detail of using horses for work. Sometimes that involves delving into technicalities which some readers may find too much (if so, just skip ahead!) but for people who need it, this sort of information can have a profound effect on the work that horses can do, and by extension, influence their efficacy as a working animal.

For many people today, the concept that draught horses can be effective and efficient may seem absurd. However, we are now in a very different position from twenty or thirty years ago as we face the unfortunate realities of climate change, with increasingly severe heat waves, droughts and floods, alongside a marked loss of biodiversity and the depletion of soil carbon. In this new reality, the role of the working horse, a power source that does not require fossil fuels and which is light on the land, should command more serious attention.

For an individual, doing useful work with horses can be an immensely absorbing and rewarding occupation. In my own life, I have spent many of my best times with horses. Any ease I have attained with this craft over that time, however, has been based on a thorough foundation, because in the beginning I was lucky enough and single-minded enough to get to learn from experienced horsemen and to work with good horses. That early training and my subsequent experience compel me to add that although there are some 'how to' aspects to this book, it is not a comprehensive primer on how to work with horses. There are other good books that aim to help with that process, but truthfully, books and videos should only ever be an adjunct to getting good hands-on instruction from someone who really knows this work.

I do, however, hope this book provides a useful insight into some of the realities, and the potential of using horses for work, whether you ever intend to harness a horse or not, and for anyone with some practical experience, a route to a deeper understanding of the working horse.

Opposite - Hampshire farmer Robert Sampson using six horses to disc the stubble.

Introduction

When we think about working horses, most of us probably imagine the past, but for me, as for others who work their horses, my thoughts are usually engaged with the present, when I am working my horses or thinking about the work we are about to do. This book reflects that reality; it is primarily about the practical aspects of working horses, about how and why we might want to use them and the benefit of using horses, for the land, for the wider environment, for individuals and society.

To use horses for work today is significantly different from how it was in the past, when working horses were everywhere, equipment was readily available, when in every village and on every city street there were people who knew how to use them. They also knew what horses could do and how they fitted in with daily life, in farming and in commerce. Today, that knowledge is no longer common knowledge; there are fewer people with refined horse skills and a detailed understanding of horse work, whilst our current economics based on doing things quickly by burning oil discourages the use of horses.

At the same time, using horses is essentially the same as it ever was, because horses are fundamentally the same, in their behaviour, their needs and capabilities. For anyone who wants to work horses, the primary qualification therefore continues to be an understanding of horses, knowing how to respond in different situations and how to guide, modify, and pre-empt the horses' instinctual reactions to enable them to feel secure whilst accomplishing the task in hand. With that essential layer of competence in place, the possibilities of what you can do with horses are both wide, and within each activity, quite specific. Those specific skills are often the hardest to acquire, because for every hundred people who can tell you how to feed and care for a horse, there are a mere handful who can fit a collar and only one or two who can explain how to adjust a plough or make hay. Because of their detailed knowledge, the experience of those who do their work with horses is therefore of great importance. Equally important is their approach, because relying on horses fundamentally changes our attitudes, to the horses, to the work, to our routines and expectations. In that process, the horse is transformed from a companion animal, a field ornament or a page full of pedigree, to a work mate and partner.

To understand the role and the value of the draught horse, he must be seen in that setting, within the context of his work. For practical horsemen and -women, that means mastering the skills necessary to do the work, but beyond that, whether for the practitioner or interested observer, we need to appreciate how everything fits together, how the horsey stuff interrelates with the technical aspects of implement choice or adjustment, how our chosen area of work is limited or enriched by using horses and how their use affects the environment, our economy and our thinking.

Tricia Hills and a pair of Percherons using a modern ballast roller behind a hitch cart.

Chapter One
A Farm Powered by Horses

The original version first appeared in the Heavy Horse World in 2003.

In the 1970s and 80s, Geoffrey Morton was one of the most recognised farmers in Britain. His fame amongst the general public was mostly due to the film made by Frank Citanovich, with the slowed-down title, 'Beauty, Bonny, Daisy, Violet, Grace and Geoffrey Morton.' In 1975, the film won a BAFTA and the Prix Italia, the combination of beautiful camera work and Geoff's measured delivery as he described his life working the farm with horses making 'Beauty, Bonny...' a special film, encapsulating a way of life which seemed timeless.

When the film was made, Geoff's way of farming was already thirty years out of date, everyone else having switched to tractors, but amongst the local farming community he was respected, if not always understood, for his hard work and tenacity. Enthusiasts for the heavy horse, meanwhile, appreciated his knowledge and commitment to keeping the tradition alive, though there were some, invariably those who had given up on horses completely, who thought his methods were not traditional enough! By the time I met Geoff in the late 1980s, he was also running training courses with his sons Andrew and Mark, and it was as a course participant that I first became involved with the farm.

On first impression, Hasholme Carr Farm did seem very traditional, centred around a square, brick-built fold yard, where along two sides and under the cover of open-fronted sheds, horses were tied up at long mangers. Along the other two sides of the yard ran a concrete path, past the door to the barn where the feed and extra harness was kept, and round the corner to five more doors. The first of these was the stable, separated into individual stalls by strong wooden partitions, with various pieces of harness hanging on the stout end posts and on the rear wall. Two of the stalls were empty; in the other two, a pair of big black horses stood contentedly, their huge rear ends towards us as we peered round the door.

On leaving the fold yard, we followed Geoff past the implement sheds stuffed full of strange and sometimes rusty pieces of machinery, down a track, past a small wood and out into one of the fields. Being springtime, Andrew was topdressing the wheat with granular fertiliser, using the pair of horses that was missing from the stable. The two mares were attached to a green-painted vehicle which consisted of nothing more than a platform, a seat and a pair of wheels, with a pole going between the horses to allow it to be steered. This hitch cart, as Geoff called it, enabled the horses to tow tractor equipment, in this case a modern fertiliser spreader, which used the turning of its wheels, via belts and pulleys, to move the fertiliser to the back of the spreader where it was flung out by a spinning disc powered by a small engine of the size you get on a domestic lawn mower. If we think about horses working on the land, this is probably not the image we have in our minds, but as Geoff explained, this was one solution which allowed them to use their horses effectively, using modern equipment rather than having to rely on old and often decrepit machines.

Andrew Morton with the Tasker Patterson fertiliser spreader.

When we returned to the stable, the practical part of the course started with the basics, how to lead a horse, tie it up securely, and how the harness goes on. Then we practised driving a horse around the fold yard, learning the words for left and right, before hitching the two good black geldings to a set of chain harrows and taking turns to drive up and down the field, a task that looked easy but which demanded continuous attention. By the end of the course, we had learnt how to put a horse in a cart, pulled some logs out of the wood, driven a pair in the hitch cart, and gained a small insight into the world of the working horse.

After the course, I was lucky enough to spend many days, and later, many months working on the farm. Like most farms, the work was structured by routine, to ensure the animals' needs were met whilst organising the work in an efficient manner. But it was not just our routine that was important, the horses' established sequence of expectations throughout the day led to them being content and easily handled, their routine being a cornerstone of the farm's success.

In the summer, the horses spent most of their time in one of the paddocks, handily located near the farm buildings. Each morning, the first job was to bring in the horses that would be at work that day, to put on their halters and lead them to the gate. When through the gate, we'd usually flip their halter shanks over their necks, and most would walk to the stable unguided. Once they had made their way into their own stalls and had started to eat, we followed them in, threaded the halter shank through a ring in the manger, then through a block of wood which would rise and fall as the horse moved backwards or forwards, before securing it with a knot. If we needed the horses soon, we'd pop the collars on before tying them up, slide the back band over their backs and hook the traces onto the hame hooks before leaving them to their breakfast.

Once we were ready to go to the field, the bridles went on and the cords were tied to the bit rings. Then the less experienced horses were carefully backed out of their standings, making sure they didn't catch any harness as they went through the door, but the older horses would come out on their own, edging backwards until they reached the back wall before turning slowly and making their way out to the trough in the yard. Most would take a drink and then we fastened them together, the halter shanks tied back to their partner's hame ring, the coupling band tied bit to bit, before the cords were uncoiled from the top of the hames and we drove them to the field.

In late autumn, winter and spring, the routine was different. Instead of being out at grass, the four main horses spent their days in the stable unless they were out working, which was on most days. Another couple were similarly housed, the rest being tied up under the covered sheds where they were fed two or three times a day, depending on how much work they were doing. At nighttime, all the horses were turned loose in the fold yard after we had unrolled a round bale or two of wheat straw for bedding and moved the ring feeder over a fresh bale of barley straw, which was their main source of roughage. Over the course of the winter, the level of muck in the fold yard gradually increased as the horses' food entered the next stage of nutrient recycling, the urine and manure being covered with more straw as successive bales were rolled out. The last job each evening was to wheel in a couple of barrow loads of mangolds, which the horses eagerly crunched away on as soon as they were turned loose.

Swale with a small cart full of mangolds.

The horseman's view when harrowing, showing how the eveners are lifted off the ground as the horses move forward and how the longest evener, the three-horse baulk, allows the pull of the three horses to be equalised, the horse on the right having twice the leverage as the pair on the left. The horse in the middle is a Shire, with a 3/4 Shire/1/4Thoroughbred cross on the offside, and an Ardennes/Shire cross on the nearside.

Rolling a spring-sown crop of oats with a gang of three rollers. On this flat land the drawbar embeds itself into the ground when the horses come to a halt.

Except in the most extreme weather, the horses stayed in the open rather than sheltering under the sheds, but when we arrived each morning, the older horses were waiting near the stable door and filed in, one after another, into their appointed stall. The others were caught, their halters put on and were led to their places, but most learned to come on their own, just one stage in their acclimatisation to the routine of a working horse.

In the short days of winter the work was steady, but as the days drew out the horses' workload increased as we prepared the ground for sowing beans, spring oats and mangolds. In March, as the horses were shedding their winter coats, and again in the autumn, it was not unusual for eight or nine horses to be out working in the fields, with three or four cultivating or rolling, three horses pulling the grain drill and a pair following with a set of light seed harrows.

When walking behind harrows or using a cultivator, the scene at Hasholme Carr could have been from a century or two earlier, but when using a gang of three rollers or cutting the corn with the rubber-tyred binder, we were employing post-war practices, and with other implements which were attached to the hitch cart, we were using an American innovation of the late twentieth century. This mixture of traditional and novel methods allowed the Morton family

This Danish-made JF binder has a five-foot cut, and unlike the earlier binders which have a big steel drive wheel, the JF has a small rubber-tyred bull wheel to drive the mechanism. Due to the small size of this wheel, no elevator canvasses are needed, so the corn goes straight from the bed into the packers. This reduced mechanism, combined with light construction and narrow width of cut means that it can easily be pulled by two horses, though in the heat of harvest, the horses were swapped for another pair after a few hours.

Two Shires and two Clydesdales on a Flexicoil Land Packer, a locally made implement which consolidates the soil to a greater depth than a Cambridge roller, especially on this peaty land.

to make the most of the horses' advantages, the lower costs, their flexibility and their light imprint on the land, incorporating time-honoured horse skills with newer machinery, some of which had not been invented when horses were in widespread use.

When Geoff started farming in the 1950s, however, he had little choice but to use what was readily available, both as regards the horses and the machinery. Here in the East Riding of Yorkshire, it was the Shire, or to be more precise, a horse of Shire type which predominated, and for machinery, there was still plenty of second-hand horse-drawn equipment going cheap at farm sales.

For ploughing, Geoff started out using traditional single-furrow ploughs, he and his father often going to plough together, each with a pair of horses using Ransome LCP ploughs or a wooden-beamed plough made by Edlington of Gainsborough. The two models were very similar, so when working together one could follow the other on the same piece of land, each ploughing a 12-inch *(30cm)* furrow typical of a semi-digger, rather than the more common nine inches *(23cm)* of a general-purpose plough, already a time and energy saving solution to a time-consuming job. As the horses came out of the yard in the mornings, each man would take whichever horses he came to first, so sometimes the pairs would be the other way round from the day before, or perhaps Geoff's nearside horse from yesterday would now be his father's offside horse.

The versatility of the horses in being yoked in different configurations continued throughout the year, not just when working as a pair but also in teams of three or four which were often used on harrows, cultivators and rollers, a practice which had been common enough on larger farms in the area. Versatility remained the key to working the horses successfully; with the horses in full work and doing every task on the farm throughout the year, they were accustomed to life in harness, willing and easy to manage, a reflection of the skill and calm approach of those who were driving. As younger horses got to an age to start work, there were always reliable older horses to yoke them with, the youngster soon picking up clues from their elders and eventually becoming accomplished work horses themselves. With an increasing number of horses, Geoff often had horses to sell, some going to other farms, others to the breweries, and whilst still on the farm there were always plenty to choose from, enabling more horses to be used at once and often allowing the mares with foals at foot to be excused work until autumn.

Watering the horses before going to the harvest field.

Unlike the horses which grew their own replacements, by the 1960s there was no longer any horse-drawn equipment being made. A farm sale might sometimes turn up a useable implement, but increasingly Geoff turned to machinery designed to be used with tractors. Some, such as harrows and trailed cultivators, were almost identical to the horse-drawn models and could be hooked directly onto a set of eveners for the horses to pull, but wheeled implements, which in a horse-drawn version would have shafts or a pole for steering and braking, required a different solution. This is where the hitch cart came in, the pole and eveners on this one piece of equipment providing the means to attach the horses, whilst the hitch at the back allowed many different implements to be used by just pulling the pin and hitching to something else.

What was happening in front of the hitch cart, or any other implement, however, was essentially the same, the horses' capacity for work remaining unchanged. So, although using the hitch cart with tractor equipment allowed the Mortons to continue to use their horses, it did not necessarily make the work any faster. In fact, tractor equipment tends to slow things down because it is more heavily built to withstand the greater forces placed on it, making it harder to pull than horse-drawn machinery. On the flat land of Hasholme Carr, this extra weight was less of an issue than in hilly country, but sometimes an extra horse would be needed, so whereas a pair was enough to pull a thirteen-coulter horse drill, the heavier fifteen-coulter tractor drill behind the hitch cart required three horses.

Mark Morton driving three horses from the hitch cart, with a Massey Ferguson no 29 drill, the last of the tractor models to use its wheels to drive the mechanism.

Because they had plenty of horses, adding an extra horse to a team made the transition to tractor machinery relatively easy and allowed each horse to work less hard as the load was divided between more horses, speeding up the pace of work as the horses could walk a little faster and needed to be rested less often. The other method of increasing the work rate was to deliberately choose wider implements to be used with bigger teams of horses.

Although Geoff had used teams of three and four horses from the start, he also knew that bigger teams had been employed in America. Then, at the Shire Horse Show in 1970, he got talking to a horse farmer from Wisconsin, who found him a second-hand two-furrow gang plough which he shipped across the Atlantic. Along with the plough was a booklet on how to hitch bigger teams of horses, in two or three rows, including the measurements for the compensatory eveners which allowed each horse to pull an equal share of the load. This innovation not only enabled Geoff to increase the output when ploughing but to use bigger teams on rollers, harrows and cultivators, and in doing so, pioneered the use of big teams in Britain.

From then on, big teams became a frequent sight on the farm, sometimes up to eight or nine horses, often pulling two implements at once, such as a cultivator or a set of rollers with harrows behind. A team of six, however, was more common and became the standard hitch for the gang plough. On light land the plough could in fact be handled by four fit horses, but with two 14-inch mouldboards, a team of five horses was more appropriate, but six were easier to drive, each of the three wheelers having a leader to follow, and the extra power meant that breaks for rest were reduced. With plenty of power, a narrow packer was often pulled to the side and to the rear of the plough, which consolidated the ground ploughed on the previous round, not only accomplishing another job but making the soil easier for the horses to walk on during subsequent operations.

Geoff Morton driving a six-horse team from the hitch cart with a flexicoil and a set of harrows behind.

When driving big teams using the Talkington method of compensating eveners, only the leaders are driven with reins. The load between the leaders is equalised in the usual manner, their power being transferred along the lead chain to the rear set of eveners, where their pull is compensated against the wheelers using eveners of varying proportions. The rear set of horses, the wheelers, are not driven by reins, but each horse is equipped with a buck back rope which clips onto each side of the bit and runs through the hame ring before joining together and continuing down the horse's side where it is clipped into the lead chain. If a horse is not next to the lead chain, its buck rope is clipped into the trace of its neighbour. When the leaders slacken their pace or stop, the lead chain becomes slack, putting pressure on the wheelers' bits. To ensure the wheelers move ahead, they have halters under their bridles, into which a light chain is clipped, the other end being clipped onto the trace chain or swingletree of the middle leader.

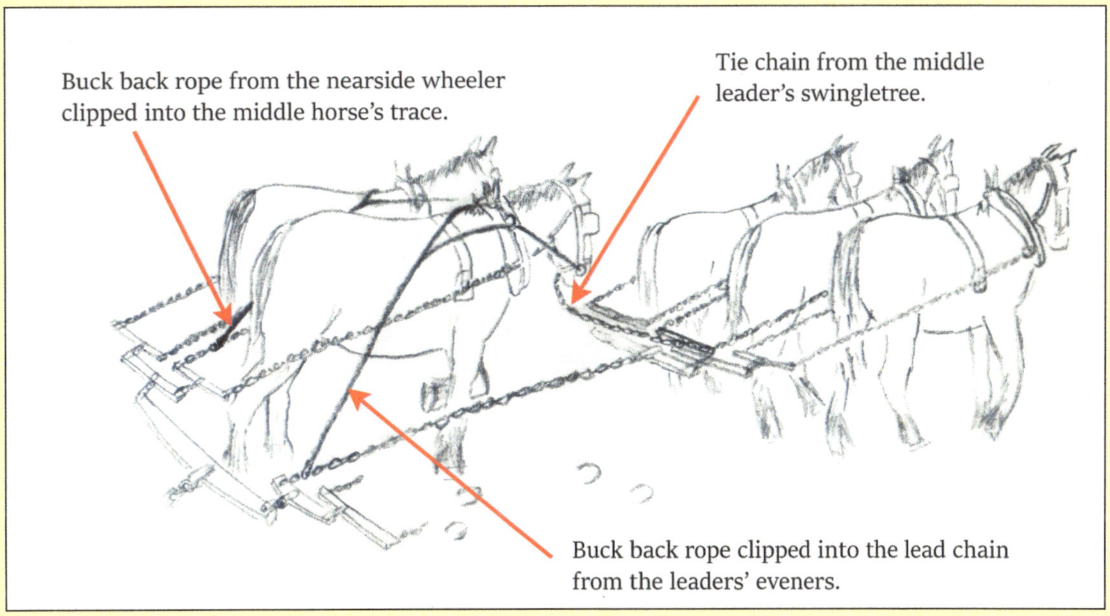

With the nearest wheeler removed from the picture (leaving only his hoofprints behind), the lead chain and the middle wheeler's buck back rope and tie chain are visible.

Throughout this time, the changing economics of farming, the increase in costs combined with a drop in prices for agricultural produce, led to another change on the farm. Until the late 'seventies, Geoff followed a traditional rotation, growing an acreage of potatoes and a one-year grass ley in between cereal crops, the oats being fed to the horses and the wheat or barley to pigs, but the changes in farm finances required a different approach. Because the Mortons were already harvesting with a binder, the main change was to increase the acreage of wheat and rye, and to sell the unbroken straw for thatching houses, making the most of the horse-drawn technology to produce a specialised and higher value product.

Of all the pre-motorised activities on a farm, harvesting grain with a binder in the traditional manner is one of the most time-consuming. Compared with a combine harvester which cuts the crop, threshes out the grain and leaves the broken straw, the leaf and the glumes behind in the field, the binder only cuts the crop and throws it out in bundles. These sheaves, tied together with a piece of twine by the binder's mechanism, then need to be stood up by hand, a pair at a time, followed by more pairs to make a stook of usually ten sheaves. Once they have stood in the field for at least two weeks to dry, they are forked onto a vehicle, taken to the yard and stacked until the winter when the sheaves are put through the threshing machine to release the grain and to produce the straight unbroken straw suitable for thatching.

Loading sheaves of wheat onto a rulley attached to the back of the hitch cart.

By growing thatching straw, the Mortons were making an economic asset out of a horse-powered necessity. Although early combines in America were ground driven and pulled by teams of up to thirty horses, in Britain combines arrived after the tractor. Many binders were therefore pulled by tractors, but using horses does have advantages. The horses may be slower, but they can turn more quickly at the corners, and, being attuned to the pull of the binder, if the machine jams up, they stop before it breaks. Using a binder with a tractor also needs two people, one to drive the tractor and one to operate the binder, whereas the horse-drawn binder can be operated by one man, freeing up another pair of hands for stooking the corn.

Andrew Morton with a rulley load of oats.

Once the sheaves were dry enough to be gathered, the horses were again in their element, moving from stook to stook on voice command or being driven by a child, whilst one or two forkers forked the sheaves up onto the load. The increase in autumn-sown crops, however, had the knock-on effect of moving more work to the autumn, making it necessary to use a tractor to get though the autumn ploughing. For horse-drawn purists that was perhaps a shame, but at least the tractor was being used at a time of year when the ground was dry and the soil hard, minimising the damaging effect of its weight. By getting ahead with the ploughing, however, the subsequent cultivations could be done by the horses, at times when their light imprint on the soil was more important.

Away from the farm, there were more changes afoot, not just within farming but with the horses themselves. When Geoff started out, most draught horses in England, excluding East Anglia and the northern counties of Cumberland, Northumberland and Durham, were more or less of Shire type. Thirty years later, the remaining horses were mostly pure-bred, pedigree Shires, most being kept to compete at shows, and increasingly, they were being bred to be popular with show judges rather than being chosen for their working abilities. Whilst the Mortons' horses were busy working in the fields, these changes were of little importance, but when it came to breeding the mares, the stallions hired by the local hiring society increasingly tended to throw foals which were taller and narrower, less suited to draught work. This was one reason why Geoff went to France in the 1970s and bought two Ardennes mares, a shorter and more compact animal than the Shire, to help establish the breed in this country and which he thought might better suit beginners. A little later he also acquired a Clydesdale mare and bred a few foals, then swapped a couple of Shire foals for a Percheron mare, so by the late 'eighties, there were Shires, Clydes, Ardennes, Percherons, and some cross-bred animals at work on the farm. In subsequent years, the balance of breeds continued to change, so as the Shires were gradually sold, most of the remaining horses were Percherons.

Although Geoff recognised that all the draught breeds had a place, as he told me during a long and wide-ranging conversation in 2003, he had been a long-standing admirer of the Percheron; "I would have had Percherons right from the word go if I could've done. In fact, the first stallion we ever used on a mare was a Percheron stallion, but we never got any foals. So, then there was no question of Percherons at all, there weren't any up here, we had no transport, and we couldn't afford to send mares away. It was only much later we swapped Folley for a couple of colt foals, but it wasn't a particular decision to change over, it's just that the Shires got more extreme all the time."

Of all the changes on the Mortons' farm, Geoff considered the use of big teams as the most important, the multiple hitch with all the horses equalised against each other allowing more horses to be worked together and more work to be done. Six horse teams became very common, and eight horse teams too, though Geoff added, "we used eight because we had enough horses available to do that, so you might as well, and get a bit more power. For a farm this size, fives and sixes was about right." However, he saw no merit in using very big teams just for the sake of it; "the principle behind the big team idea was however many horses you used to work your farm, for some of the time, you organised it so you could put them all into one team, rather than have them all split up into pairs or whatever. It doesn't mean to say, if you've only got three horses, you've got to go and get yourself another two or three and make a five or six-horse hitch, it means make sure you always use the three together when you can. And if you have got four, five or six, well make it so you can use them all together; it's a fundamental part of it. So is the business of driving the leaders, with buck ropes and tie chains on the wheelers."

Although driving multiple hitches seems complicated, the horses work well in big teams. Indeed, Geoff often put a young horse in the nearside wheel position of a six-horse hitch after they had been yoked a few times as a pair alongside an older horse. Not only had they no

Percheron mares and foals.

option but to move forward after the leaders and stop with the others, but as a herd animal, they had the security of working quietly alongside their fellows. Geoff also found this hitch easier to drive than many horses yoked abreast and it was easier on the horses when turning. It did, however, require a different way of driving because without having separate reins for the leaders and wheelers or a separate point of attachment for their eveners, you cannot hold the leaders back and take them out of draught as when driving four in hand. Also, there isn't much bend in the hitch, meaning the horses must sweep round the turns, which of course requires wider headlands than normal.

Because of his wide experience with working horses, Geoff understood there were wider applications for horses to be sensibly employed. One area he cited as having a largely untapped potential was on smaller organic farms, though for anyone wanting to use horses, he sounded a note of caution; "everything to do with horses tends to be very complicated. All this really depends on you knowing how to use these implements, but much more than that, it depends on how good you are with the horses and how good the horses are. You can't do any of these things unless you're something like a reasonable horseman.... It's absolutely useless saying horses are fine and efficient and very helpful, especially for weed control in organic systems, unless it's to somebody who understands horses. It's the same problem for everybody, but if you're real good at driving horses, almost without thinking, you've a spare little bit of capacity to look at all the other things."

"When you're doing anything with a horse and things aren't going right, it's knowing whether it's the horse, the harness, the implement or you. You maybe think, 'oh it's that, that's wrong,' and you put it right and you're maybe correct in your thinking. You alter it to the way it should be, try again and it's still doesn't go right because it was not only that, but it was the horse as well or you as well. So you put it back where it was and alter something else, and you still haven't got it right. Once you get more than two things to consider there's an awful lot of permutations."

Despite the challenges, Geoff remained enthusiastic about the future of the draught horse, whether used for real work, for work as pleasure, or even for riding. His principal interest, however, remained in the use of the horse as a draught animal, and not surprisingly, he viewed farming with horses as an entirely valid way of working for anyone who wanted to try it on a proper farming scale, and one which could well return in the future. His lifelong passion for the working horse would be ample explanation for wanting to see it continue, but typically, he had an entirely practical reason; "I hope somebody will keep this going, because knowledge once you've lost it, is very difficult to get again."

Geoff making potato rows in the 1970s.

Opposite - Jonathan Waterer turning hay with a New Holland side-delivery rake behind a hitch cart.

Chapter Two

Two more Farms

Heavy Horse World, 2003 and 2005

This chapter comes from a series of articles about farmers who use horses to do their work. The focus was on how they faced the challenges of finding suitable machinery to use with their horses, but inevitably their choice of equipment was inextricably linked to their attitudes and approach to farming, the land and crops, their horses and horsemanship.

Higher Biddacott Farm

After driving along the quiet enclosed lanes which wind their way up and down the hills of North Devon, you might think that nothing much happens here. But Higher Biddacott Farm, the home of Jonathan and Fiona Waterer, is a hive of purposeful activity, with horses coming and going, dogs running about, and people arriving for Bed and Breakfast or to enjoy the Devon landscape from the back of a horse-drawn waggon. At Higher Biddacott, working horses are at the centre of everything; besides the waggon rides, the horses are used for weddings and funerals, to teach people to drive and to work the farm. In addition to working their own horses, Jonathan also trains horses for other people, aiming to send away two trained horses each month. Important as these activities are, Jonathan is primarily a farmer, committed to farming as best he can. Although the outside work with the horses provides most of the income and allows him to farm in the way he wants to, the farm in turn facilitates the other enterprises by providing regular work for the horses, which keeps them fit, versatile and dependable.

Although Jonathan was brought up on a farm, his approach to working horses was greatly influenced by the four years he spent in western Canada as a young man, working on a ranch with Appaloosa and Percheron horses. Whilst there, he had the opportunity to learn about training and breaking horses; "I was privileged enough to end up working with more horses, 'specially heavy horses, than probably most people, and in every sort of task, so from that point of view I was terribly lucky, as the more you do, the easier it all is. You cannot beat working with lots of different horses."

This experience stood him in good stead, as he continued to train horses for other people when he started out on his own place in 1982. This was on the edge of Exmoor, where he managed to buy a very hilly farm of 44 acres, *(17 ha)* which did not have a house until Jonathan converted the old cow shed. Then in 1996, he and Fiona moved with their children to Higher Biddacott, a farm of 87 acres, *(35 ha)* most of which is under grass, but with seven acres *(3 ha)* sown to wheat grown to produce wheat reed for thatching, and another seven acres of woodland. All the farm is hilly, with a stony loam soil, and typically for the West Country, it receives forty-five inches *(1150 mm)* of rain a year. The Waterers have seven horses, one mare and six geldings, all Shires. They are usually bought in as youngsters as Jonathan finds them easy enough to buy, and to some extent you can tell what you're getting without having the risk of breeding. There are also forty head of cattle, 'ruby red' Devons, which are put to a South Devon bull to give bigger calves. Until a couple of years ago they rented some additional land and had more cattle, but reducing the numbers has simplified the system, freeing time for training horses, especially in winter. Now all the meat from the cattle is sold direct to the customer after being slaughtered locally and is delivered in quarters of mixed cuts.

Although every horse which comes to be trained must be accustomed to work in shafts, cart harness is only occasionally used on the farm, for odd carting jobs and on the trip rake. For the rest of the work, Jonathan uses two or more horses and North American team harness. An important feature of this Western Breeching style of harness are the quarter straps from each side of the britchen, which go under the horse's belly and clip onto a pole strap which passes between the horse's front legs and clips onto the neck yoke. This transfers the force of the horse pushing backwards in a straight line from the britchen to the neck yoke without interfering with the collar, allowing the vehicle to be reversed accurately and the horses' efforts to be effectively transferred to the vehicle. When Jonathan returned from Canada, he brought a set of harness hardware with him and had a set of harness made, which is still in use twenty years later. As such, he was probably the first to use North American harness in this country in recent years, and he still strongly advocates its use, describing the britchen set-up as, "the most efficient system of backing in the world. When I came to Devon, I knew I had to have something which would hold vehicles back on the hills." He is also keen on the leather traces, hating to see sores on horses' legs caused by chafing of chains, and he appreciates the use of collar pads under the collars as shoulder sores are very unusual when they are used.

Comfortable horses with collar pads and Western Breeching harness.

Walking round the farm, there is little evidence of any equipment at all as Jonathan can't bear to see it sitting outside in the rain, so until a planned implement shed is built, it is all crammed into whatever shed space is available. The machinery is a mix of old horse-drawn tools, some tractor implements and several new or reconditioned horse-drawn machines, all of which are obviously well looked after.

The first new implement Jonathan bought was a ride-on sulky plough, made by Pioneer Equipment in Ohio, which is used with three or four horses. It is a frameless model, so the seat and wheels are attached directly to the beam. It has a disc coulter in front of the plough body with a fourteen-inch *(36cm)* semi-digging mouldboard, and a cover board on top to turn the stubble under. The two wheels are raised and lowered with levers, with a third lever to angle the furrow wheel to direct the plough towards or away from the land, or to allow the wheel to caster when turning on the headlands. Jonathan finds it very easy to operate; "it's all on levers, there's no fiddling about with a spanner, letting wheels up and down, and as you go along

Jonathan ploughing grass in preparation for a wheat crop using his Pioneer sulky plough.

The Thomas and Son corn drill, still in use.

you can soon alter something. And, of course, the main thing is you can write to the Amish and they send the parts to you, instead of wandering around at farm sales, wondering if you can get another share somewhere." The hardest part of using it, particularly on hilly land, is learning exactly where the share needs to go when getting back into work again, as there are no handles to control it, but Jonathan describes it as, "a wonderful plough. You whip along and can certainly see some work being done."

After ploughing, the land is worked down with a set of harrows which belonged to Jonathan's father, and rolled with a single, wide section of Cambridge roller behind the hitch cart. The corn drill, by contrast, is an old horse-drawn implement, made at least fifty years ago by C H Thomas & Son of Bishops Tawton, which Jonathan chanced upon at a local sale and bought for £30. It has steel wheels, force feed to 15 plain coulters and covers 6 feet *(1.8m)* on each pass, which is ample width for a pair of horses on these hilly fields.

Not being great believers in the use of sprays, the next operation in the cornfield is cutting the wheat. The binder is an International, which Jonathan bought at a farm sale when he was fourteen! Most of the binders he saw working in Canada were Internationals, with a 6 ft or 6'6" cut, *(1.8 to 2m)* pulled by four horses abreast, whereas Jonathan's is a 5'6" *(1.7m)* model pulled by three horses. The horses go straight on the pole as was usual in Britain, though he would like to try a tongue truck, a pair of small wheels which take the weight of the front of the binder. Jonathan likes tongue trucks on mowers as well as binders because it also takes the weight off the horses' necks, but warns against using them with inexperienced horses as it is possible for the horses to pivot right round and onto the cutter bar.

Once the sheaves are dry, a big bale trailer is used behind the hitch cart to carry the corn to a stack in the field until the autumn when a contractor comes with a reed comber to produce the combed wheat reed used for the distinctive West Country style of thatch.

Besides rolling and carting sheaves, the hitch cart is also used on the side-delivery rake, occasionally for chain harrowing, and for spreading fertiliser. The Waterers do use a little artificial fertiliser on the grassland, except for the land which is in the Countryside Stewardship scheme and another ten acres which is sown to a mixture containing a high proportion of clover. The remaining grass gets about a bag per acre, spread with a ground-drive Wessex spreader made for an ATV, which has a 12 foot *(3.6m)* spread.

The Waterers' second piece of brand-new horse-drawn equipment is a Canadian manure spreader, which they imported a few years ago from Jonas Kuepfer, an Amish farmer in Ontario. Although costing £2000, Jonathan sees it as a good investment; "it works incredibly well, and kept undercover it should last my lifetime." It is loaded with a tractor bucket and is pulled by two horses, except on the steepest ground when four are used.

Also from Ontario is Jonathan's mower. It is a McCormick Deering (International) No 7, which was bought after being completely reconditioned in Jonas Kuepfer's workshop. This model was made in Chicago between 1929 and 1939, both for the home market and for export, the ones in Britain mostly having a 4'6" *(1.4m)* cut, as does Jonathan's original mower, which is now fairly worn. The new one, however, which has new bearings, oil seals, gears and cutter bar parts, cuts a 6-foot *(1.8m)* swath, enabling an acre to be cut in an hour. Jonathan much prefers this mower to the heavier English machines, a major advantage being that a complete range of new parts is available. The light weight of this mower and the enclosed gears in an oil bath also contribute to making it easy for the horses to pull, as does its 'as new' condition. Nonetheless, it is still hard work on the hills, so mowing is usually done in the cool of the early morning or in the evening, leaving the middle of the day for turning the hay.

To move the hay Jonathan has a reel-type tedder made by Nicholson of Newark. As a tractor model, it is somewhat heavier and wider than the horse-drawn models which were made for a single horse, but it works well behind a pair of horses, a pole and seat having been added to convert it to horse use. If the weather is favourable, it is used on two consecutive days before the hay is rowed up with the hitch cart and side-delivery rake. This is also a tractor model, made by Dening of Chard, but it is lightly built, raking 6 feet *(1.8m)* on each pass.

At the moment, baling is a tractor job, most of it being put into small bales, though sometimes haylage is made, for weed control and on the steepest ground. The hay is baled with a John Deere baler with a Cook sledge behind, which leaves the bales in groups of six. Raking the fields after baling is one of the rare single horse jobs on the farm. At a glance, the horse-drawn trip rake made by the Devon firm of Huxtable, looks like any other. Its working width is 10 feet *(3m)*, but when the raking is finished, its legs are lowered to the ground and the whole thing can be wound together with a handle, so it is narrow enough to negotiate the Devon lanes.

The Waterers make forty acres of hay, at a time of year which is busy with weddings and waggon rides, which means Jonathan and Fiona are working flat out in June and July. With mowing at both ends of the day, Jonathan finds that the Nicholson tedder is not quick enough when there are ten acres of hay on the ground. Although he has enough horses, he finds he is simply spending too much time on the seat of the tedder, so he has recently bought a Farr Centipede, which he plans to use behind a hitch cart with a mounted engine. Although primarily intended for tedding, the planned hitch cart with a 25 hp engine will also be powerful enough to run the baler. In tractor terms this seems low-powered, but of course the horses will be providing the motive power, and without a tractor, there won't be the lump of a tractor itself to be moved. Jonathan thinks this should make a big difference to the speed of turning hay. "I think two horses will pull it very well, but if we want to keep going, I don't mind if we put four on it; we've plenty of horses, and you would never stop all day. And again, with the baler, I would probably keep the Cook sledge behind; four horses would run that." However, "if we were only cutting twenty acres of hay, I wouldn't think of doing any of this."

Higher Biddacott, updated

Over the last twenty years, some things have changed at Higher Biddacott. Jonathan's intention to have a hitch cart with an engine came to fruition with the arrival from America of a Pioneer hitch cart with a 43 hp diesel engine, which took a whole day to unpack, but it still baled a hundred bales of straw once it was hooked up to the horses the same evening. As when baling with a tractor, Jonathan uses the bale sledge behind the baler, so uses three horses on the hitch cart, the same configuration usually being used on the Fahr hay tedder.

To increase the speed of mowing, Jonathan also uses the powercart with a Claas drum mower which, with its 8ft 6-inch *(2.5 m)* cut, can cut five acres in three hours. Unlike most modern mowers which are mounted on a tractor's hydraulic arms, this is a trailed model which is lifted on its own wheels to take it out of work, avoiding the balance issues associated with using a mounted implement on a hitch cart, the hitch cart being equipped with spool valves to operate the hydraulic cylinders on the mower.

The adjustable link on the powercart.

A special feature of the Pioneer powercart is its additional strut which fixes the position of the implement in relation to the hitch cart. This consists of a hydraulically adjustable link which is secured to a pin on the implement's drawbar, preventing any horizontal or vertical movement between the two. Once the hitch cart wheels have been unlocked so they are now steered by the pole, the whole set-up functions as a four-wheeled vehicle. This improves its stability and allows the PTO shaft to always run straight whilst preventing oversharp turns which could otherwise result in the horses getting too near the machine.

Another change came in 2016 when Jonathan bought two Swedish-built, Activ combines, a 1960s model designed to go behind a tractor. One went to a friend in France who used it behind a Pioneer forecart with a lighter 35 hp petrol engine, whilst Jonathan kept the other for his own use. Jonathan had long wanted to make flour for bread making, but previously the grain from the long-strawed thatching wheat was sold for animal feed. Whilst in France, however, he learnt so much about the growing and milling processes on his friend's regenerative, no-till, organic farm that he decided to grow some at home. Now that Higher Biddacott is certified organic with the Soil Association, the grain goes away to Shipton Mill to be stone ground into flour and is sold from the farm and other outlets in one kilo *(2.2lbs)* bags or twenty-five-kilogram *(55lbs)* sacks. When using the combine and the Pioneer forecart, four horses are necessary to negotiate the hills. With only a small area of wheat to cut, Jonathan waits until the afternoon before cutting so the grain is at its driest, harvesting two or three tonnes by the end of the day.

Over the years, many different horses have been used for mowing and combining, but running powered machines does present some different challenges from ground-drive equipment. The noise of the engine is something the horses soon get used to, particularly when they can see the lack of reaction in their teammates who have done the job before, but Jonathan notes, "they need to be in full work before you put them in a team mowing or combining. There is no room for problems whilst using these types of machine." Some people also express concern that the drum mower might throw out stones towards the horses, but Jonathan has not had that experience, all the hay, and anything else, being projected backwards.

Four horses stand while hitched to the powercart with the combine behind.

The most recent piece of equipment to arrive on the farm is a Pioneer culti-mulcher, a tool now popular amongst horse farmers in North America. It has a serrated roller in front, followed by three rows of cultivator tines and a crumbler behind, effectively doing three processes at once, the forward-facing tines helping to put downwards pressure on the rollers, which in turn provide a solid frame for the tines to work against. Depending on the soil conditions and the depth of work, this tool does take some pulling, so despite its relatively narrow six-foot width, *(1.8m)* on this soil a team of six horses is usual. It is extremely effective, however, only one pass being enough to turn the ploughed ground into an autumn seedbed.

The culti-mulcher, a simple but effective machine and a great time-saver.

In 2025, as in 2003, Jonathan and Fiona's dedication to good farming with horses is evident everywhere, in their care of the horses, the harness and equipment, as well as the buildings and fields. With all the outside work they do, the farming system needs to be simple and streamlined, their policy of investing in new or reconditioned horse-drawn equipment continuing to keep the farm running efficiently. Not only do Jonathan and Fiona have the right equipment, they know very well how to use it and they obviously enjoy the work. Their commitment is impressive; their enthusiasm for the farm, the horses and their way of life is infectious. Higher Biddacott was, and still is, an inspiring place.

Lyng Farm

On the other side of the country from the Waterers is the small horse-powered farm of Joe Godderidge. Although set in the largely arable county of Norfolk, Lyng Farm is mostly under grass, and until Joe was in his late sixties, he milked twenty Jersey cows. However, when the income from small scale dairying became uneconomic, the dairy herd was replaced by a small beef suckler herd.

Joe uses horses for nearly everything on the farm; only the loading of the manure spreader is a tractor job. The horses are Shires of the older type, not too tall, which all descend from a mare Joe bought in 1974, though he already had experience working on his parents' market garden with a Percheron gelding. The usual number of work horses on the farm is five, and with many operations, all five are used in one team. The soil is heavy clay, Joe describing it as, "definitely three-horse land," a term used by old-time farmers meaning that the soil is so heavy that it requires three horses to pull a traditional single-furrow plough.

The farm's rotation is based around a grass and clover ley, which after three or four years is ploughed up for a crop of kale. This is strip grazed by the cattle before being ploughed again and sown to oats and the grass mixture. Instead of waiting until the oats are ripe, they are cut at the milky stage to make oat hay, which gets around the problem of drying grain and getting a combine small enough to cut a small acreage. This leaves the grass and clover for the next ley. Since the dairy herd went, Joe has relied on his paintings, prints and cards for more of his income. Most of them feature working horses, and his horseman's eye is evident in the way he captures the motion of the horses on canvas. Joe usually paints in the evenings, leaving the daylight hours to work outside, but inevitably this change of emphasis has affected the farm, especially as Joe and his wife are away from home most weekends, exhibiting his work at shows and ploughing matches.

Joe's painting of ploughing 'on the square.'

Unlike the teams used at ploughing matches, Joe ploughs with five horses on a Pioneer sulky plough, the same model as Jonathan Waterer's, with three horses in the lead and two wheelers behind them. Hitching the horses like this, with the wheelers spaced widely apart, gives Joe a good view of the leaders from the seat of the plough. When he had fewer horses, he did plough the whole farm with his three mares, but with five they can go all day, scarcely needing to stop for a blow, so even on this heavy land Joe can plough an acre in four hours. Although some may think it a disadvantage to have a strung-out team which needs wider headlands for turning, Joe doesn't see this as a problem. Indeed, he likes to plough the whole field as if it were a big headland, the plough never coming out of the ground at the corners. This ploughing, 'on the square' as it is called, starts at the middle of the field, where a relatively small rectangle is ploughed so that its edges are equidistant from the field boundaries, then the rest is ploughed round and round all four sides until the field is done. Although the shape of most of his fields doesn't often allow him to do this, Joe likes the economy of the system and the challenge of getting the corners right; "one of the joys of it is to discipline the horses so they stop exactly when you want them to, and they do catch on. You say 'wheesh round, wheesh round hard,' they fan round, and you just watch that plough go round and it's lovely, it's in exactly the right place, 'go on!' and away they go again. It's just such a nice experience, if they're going alright!" For the horses to work as well as this, Joe adds, "they do need to be used all the time. I don't use them as much now as I did, and I regret that, because if you get a team going like that, you can talk them round, they're lovely to use."

For the cultivations, Joe uses the five-horse team in the same configuration. Usually the discs come first, though using harrows or rolls might theoretically be better, but on this heavy soil it can be difficult to catch the ideal conditions after ploughing. The discs are a Ferguson tandem set made for a tractor, with a pair of wheels added on the front to take the weight off the drawbar, and a platform and seat at the back. Besides putting a bit more weight onto the rear gang, this is also the safest place for the driver. After discing, the land is worked down with harrows or a gang of rib rolls attached behind the hitch cart. The harrows are a 15 foot *(4.5m)* tractor set with 7" *(18cm)* curved tines, and on this land they are a good load for five horses.

Besides the plough, Joe's other new horse-drawn implement is a ground driven muck spreader, which came from Canada in the same shipment as Jonathan Waterer's. In comparison with its predecessor, a two-wheeled International made for a tractor, which required three horses and a hitch cart, the new spreader is easily handled by two horses, even on the sloping ground, as long as the muck has rotted down a bit and isn't straight out of the fold yard. The Ackerman steering on the front wheels also makes it very manoeuvrable; as Joe puts it, "you could back it right into the kitchen and under the table!"

The Canadian-built muck spreader waiting to be filled.

A feature of Joe's approach to farming with horses is his willingness to experiment with different implements and techniques. A good example is hay making. At one time Joe cut his hay with a tractor and disc mower, which was fine until it broke down. Trying to repair it made Joe question the wisdom of using complicated and fast-moving machinery, and after experimenting unsuccessfully with an old ground-drive sickle-bar mower made for a tractor, he determined to make his own mower. He wanted a four-wheeled machine that he could use with three horses, and after reading about the Busatis mowers with double reciprocating knives, which unlike the old finger mowers, do not clog, he decided to use one of these units as the basis of his mower.

The new mower uses parts from an old dumper truck, with the bigger drive wheels at the back of the mower, which, through the original gearbox, drive the cutter bar. The front axle also comes from the dumper, and like a car, the wheels are steered via track rods, though in this case they are connected to the pole. The 6-foot *(1.8m)* cutter bar hangs between the axles and is raised and lowered with a hydraulic ram, activated by a 12-volt electric pump powered by a battery, so raising and lowering the cutter bar is simply a matter of pressing a button. With three horses Joe can cut about two acres *(0.8ha)* an hour; "It works like a dream, and you can sit on there all day, without hardly getting off; it just doesn't block up. You've got the two knives working against each other, no fingers, and it just cuts through everything. It's a joy to use it."

Joe demonstrating his mower at the East of England showground.

Another example of Joe's ingenuity is his hoe, which he built after having a less than satisfactory experience with an old horse-drawn tool. "I had an old Norfolk steerage hoe, which was quite good if you've got a horse that is not too clumsy, has got a good mouth, will listen to you and doesn't need leading. I work alone, and it was hard work with that hoe, and I couldn't see why I couldn't adapt the same system so you could have a ride-on job. You could steer with your feet and drive the horses with your hands, because that's the trouble with the steerage hoe, when you pull the handles this way, you pull the mouth that way, and the horse gets out of sorts. I had seen illustrations in the American magazines of their steerage hoes, and at one time I had an Allis Chalmers tractor which had an under-slung steerage hoe and a very narrow belly, so you could see what you were doing, and it was ideal. So why couldn't I do the same thing with the horses?"

On a trip to the feed merchants, Joe noticed how the rear wheels of a forklift truck throw the whole thing round and decided to use the same idea on the new hoe, the fine steering of the back wheels being done by foot pedals whilst the front axle is steered by the pole between the horses. Between the two axles is a Ferguson toolbar, which would usually go on the back of a tractor. It is raised and lowered with a hand lever, though Joe is contemplating replacing it with a hydraulic lift like he has on the mower. The new hoe enables Joe to accomplish another job on his own, the advantage being clearly demonstrated one day when his neighbour was also out hoeing; "I felt quite chuffed about that because my neighbour had exactly the same thing (behind a tractor), but he had two people and I had only one person, and that was the same machine!"

Another unusual solution to working on his own is Joe's method of gathering the hay, which is stored loose rather than being baled. It is loaded onto a rubber-tyred four-wheeled trailer using an International green-crop loader, all drawn by three horses. The ground-drive green-crop loader is like the old-fashioned hay loader which is towed behind a waggon, the crop being pushed up an inclined plane by tines mounted to several raker bars, the hay falling on to the waggon when it gets to the top. Compared with a hay loader, the green-crop loader is more strongly built, it has a pick-up reel like a baler, and instead of a second crank at the top, it has a runner for each of the raker bars, which keeps the bars in place laterally but allows them to rise and fall depending on whether they are moving up or down.

Originally, the loader had metal wheels which made the whole thing shake and rattle especially when going along a track, but these have been replaced with pneumatic tyres which make it quieter and easier to pull. This is particularly noticeable when hitching onto the back of the trailer as there is sometimes some manhandling to be done to line it up. The tyres and wheel rims came off the retired International manure spreader but the original cast hubs were retained as they contain the pawls which transfer the drive from the wheels to the mechanism.

Down both sides of the trailer, Joe ties three lengths of rope between the hay racks which prevents the hay from falling off the sides. This is important as Joe usually works on his own. "Most of my fields are 200 yards *(180m)* from one end to the other. You start off, and I ignore the hay until it piles up at the back. By the time you get to the other end of the field there's a

mountain there, and if I stick a fork in the top and rock it, it'll fall down; I don't give myself any more work than I have to. You just have to watch at the end of the field as you go to turn. That's why I always stop at the end and pull the hay off the top of the loader, otherwise it will drop a lump off when you're going round the corner." When fully loaded, a rope to the rear ladder of the trailer is pulled to release the pin to the loader, so the loader can be left in the field.

Backing the loaded waggon ready to unload the hay into the barn.

On reaching the barn with a full load, the next labour-saving solution comes into play. This is a hay grab, which were once fairly widespread in Britain, usually suspended from a pole for building outside stacks. Joe, however, uses it with a track in the barn, a method that was common in the big barns of North America, Joe's method being a simplification of the American system. The hay grab hangs by a steel cable from an electric winch which is attached to the underside of a trolley that travels along an I-beam running the length of the barn. In use, the tines of the grab are pushed into the load, and as the grab is lifted it closes round the hay. When at the top, the trolley with its load of hay is taken to the back of the barn by pulling a rope, which goes from the trolley, through a pulley in the back of the barn and back to the operator. When the load is above the stack, it is dumped by pulling a cord to the trip mechanism. Continuing to pull on the cord returns the trolley to its starting position above the trailer. Throughout the operation, Joe stays on the load, though ideally it would be helpful to have an extra person in the barn. Nonetheless, having done the same job on his own with conventional bales, Joe is in no doubt that it is easier to cart the hay loose. Added to that, it is better hay, there is no plastic string and no complicated machinery to maintain.

For Joe, having the right equipment and good horses all contribute to the efficient running of the farm. Not surprisingly, the way he keeps his horses is also simple and effective. Although they are fed some bought-in concentrates, the mainstay of their feed is hay, which is forked down daily as the barn is next to the open yard where the horses spend the nights in winter. Not only is this an easy and economical way to keep them, but it also has benefits for the horses. "I only spend about £3 a week per horse on hard food, all the year round. Apart from that they just get hay, which is home grown, and straw. They eat a lot of straw, and I just keep them in the yard; I don't tie them up at night, and I don't keep them in loose boxes; they're best in a straw yard. There are shelters they can go in, but they hardly ever go in them. I've seen them with inches of snow on them; they're very hardy animals, they don't want shutting in."

As someone who has spent time and effort in ensuring the horses are used to their full potential, Joe knows that there could be a wider role for the farm horse. Although he is obviously happy with doing all his own work with horses, he doesn't necessarily think this should be the way for other people to go; "my own thinking is that a team of horses could replace that second tractor. I don't think anybody should think about doing everything with horses completely."

To many people, the idea of using horses to work a farm might seem set with limitations, but being a good horseman himself, Joe knows how to make the most of the horses' advantages. An important part of his success lies in his ability to come up with practical solutions, using a mixture of old and new technology, enabling him to do many jobs on his own which otherwise would require additional labour or expensive machines.

When Joe speaks about the practicalities of working horses on the farm, he effortlessly draws together the connections between the horses and the land, the crops, machinery and money. Just as his paintings come from his experience of life, his belief in using horses comes from his experience of doing the work. But, even when it comes to the bottom line, he doesn't lose sight of his basic motivation, which is fundamental to making it work; "when we were dairy farming, I don't think I would still be in possession of this place if I hadn't got the horses, because they definitely don't cost so much, you haven't got the machinery repairs and replacements. If you've got a good team of horses, there's nothing like it."

Chapter Three
Making Hay with a Single Horse

Small Farmer's Journal, 2012

When this article was first published, I had been working my own horse on small patches of land for a dozen years, making hay every year. My activities were often limited by time, money and the availability of equipment. This account is therefore particular to my situation, but at the same time it illustrates the thought processes, the challenges and some solutions to making hay using old equipment and the relatively low power of a single horse.

Making hay nowadays seems simple; get a powerful tractor, mow the grass, throw it around for a few days to dry, then put it into bales to keep for winter; simple, as long as the sun shines!

My introduction to hay making, however, was a little different. Although most jobs on Geoff Morton's farm were horse jobs, by the time I worked there, the hay was cut and baled with a tractor, so it was only the turning and tedding that were horse-drawn, both jobs being done with a pair of horses in a hitch cart pulling implements made for tractors. My own start came ten years later, and as far as implements and power were concerned, I was starting from scratch. My Percheron filly, which I bought as a weaned foal, was still too young to work, so for the first couple of years I got someone to cut and bale the hay, and borrowed a tractor to do the turning. By the time Molly was three, she had done enough work that I could use her to gather the bales, and once I had a suitable piece of equipment, to turn the hay.

Gathering bales with Molly as a three-year old.

My first hay-making implement was a Nicholson side-delivery rake, made for a tractor. It was in a corner of a field a couple of miles away, cheap, rusty and partially seized up, so I oiled it and left it for a week before borrowing a tractor early one Sunday morning and trundled it home along the road, the mechanism stuck in gear, turning the air as we went! With the application of heat and penetrating oil, hammers and spanners, I got the mechanism free again and put it to use. In its function it is identical to the horse-drawn side-delivery rakes, except that instead of the central sections of each raker bar being removable to allow two swaths to be turned independently, this rake had an additional gearbox and mechanism to turn the right-hand swath. This extra weight and gearing meant it was also heavier to pull, but it turned my few acres of hay, and in a light crop and with reasonable weather it was all that was required.

In heavy hay, however, any intertwined lumps of grass were slow to dry so I often went round the field with a fork to fluff up any clumps, especially near the hedges and at the corners where the hay turner was less effective. Noticing how much more evenly the corners dried out once I had separated the hay with a fork, I decided to look for a horse-drawn tedder. Within a year I chanced on a Bamford 'kicker' for sale. This would surely be the answer to my problems, but being at least fifty years old, it also had problems of its own.

Tedding hay with the Bamford 'kicker' in its original configuration, but with the shafts lengthened and the guard added. The swaths had been turned once and were still lying as ribbons until the tedder threw them in the air.

The first problem was that my mare only just fitted in the shafts, this implement being made in various sizes to suit horses of different stature, so even when the shoulder chains were at full length, her hocks only just cleared the cross piece joining the shafts. Even after lengthening the shafts, her tail was so near the revolving mechanism that it was at risk of getting caught in the reel, so despite the presence of flies I always had to put it into a secure plait. The other safety issue was the position of the driver's seat, just inches from the revolving tines, so before using the machine I built a guard to prevent me from toppling over sideways and ending up as part of the mechanism! Thus equipped, we went and tedded hay, though it did miss bits where some of the tines were missing. Then, whilst tedding hay a couple of years later, there was a sudden bang and we came to an abrupt halt, caused by one of the bars of the reel bending, which hit the frame and stopped the machine. In retrospect, I should have replaced those rusty and patched bars before I ever used it, but now it was a necessity. Like many repairs to old machinery, the job looked easy enough, but drilling out the extremely hard hot-riveted rivets nearly burnt out my drill and took longer than cutting and drilling all the new angle steel replacements and bolting on the tines. The biggest improvement in the function of the tedder, however, was finding and fitting enough tines to complete the set.

When using the horse-drawn tedder to fluff up the hay immediately after using the turner, it was noticeable how much easier it was for the horse to pull compared with the side-delivery rake made for a tractor. To improve the situation, I decided to make some modifications, to dispense with the hitch cart and fix on a pair of shafts, then add a seat, footrest and guard rail on the back. Having once fallen off a hitch cart which did not have a guard rail, I am a great believer in having something to stop me falling, as well as a good solid platform for my feet. Doing away with the hitch cart would have reduced the draught a little, but there was another obvious cause for the side-delivery rake's heavy draught, the small iron wheel at the front which probably carried a third of the machine's weight and which cut into ground when the conditions were anything but bone dry. This was replaced with a back wheel from a Ford Fiesta, which is twice as wide and nearly twice as high as the original wheel, and of course, has a pneumatic tyre. On this model of car, each rear wheel was fixed to a stub axle, so it was easy to bolt onto a new frame welded to the turner.

The front of the side-delivery rake in its original condition, with the replacement rubber-tyred wheel behind.

Below - The side-delivery rake equipped with shafts, the new front wheel, a sprung seat and guard rail, a more comfortable solution for all concerned.

The new wheel was an immediate improvement, and without the hitch cart it was easier to turn tight corners, I could back it up, and with a guesstimated reduction in draught of about a third, fewer rests were needed. In addition, the swaths were no longer being run over by the wheels of the hitch cart; the horse just walked between the swaths, followed by the front wheel of the turner. Another advantage was that I was now sitting at the back of the machine, so I no longer had to crane my neck round to check the turner wasn't jamming up. In the beginning this was a common occurrence, as the hay would lift and settle on top of the raker bars, eventually wrapping itself round the bearings, but now I could see this starting to happen, stop the horse and lean forward to remove the hay before it jammed. It wasn't long before Molly learned to stop of her own accord, because in an open bridle and without the hitch cart in the way, she could see what was happening. Although I could now prevent the machine from jamming, it wasn't until one year when I had much more hay to turn that the reason for the previous blockages became clear; the hay was no longer being lifted up because the tines had become smooth with wear, so the hay slid off them rather than being caught on the rusty surfaces. Since then, I put grease on the tines at the end of each season to prevent them from rusting.

A couple of years later, my attention turned again to the tedder. Although the safety guard had put me out of danger, sitting on that seat always felt precarious because the footrest was narrow and just a bit too far away, and occasionally something underneath me made a worrying

The Bamford tedder and the shafts both attached off-centre on the hitch cart.

clunking noise. Rather than trying to take the old cast iron mechanism apart, which would probably break, I replaced the shafts with a drawbar so the tedder could go behind the hitch cart, the problem now being that the wheels of the hitch cart would now run on the swaths. To remedy this, I remade the front of the hitch cart so that I can offset the shafts to one side, in line with the offside wheel. Although this looks a bit strange, the line of draught is straight down the tedder drawbar, and no hooves or wheels paddle the hay.

Using the tedder is the job where I need to drive most accurately because it was designed to work on two four-foot-six *(1.4m)* swaths. This is the cutting width of most British horse-drawn mowers, but the swaths left by the contractor's tractor mowers were five-feet *(1.5m)* wide. To overcome this problem, I always turned the hay before tedding it, which I like to do anyway to expose the damp underside of the swaths, and removed a couple of tines from the nearside of the side-delivery rake so the two rows were left closer together.

Although the addition of the shafts and the pneumatic wheel made the side-delivery rake more usable, the lack of replacement tines and other breakages eventually made me start to look for a better solution. Eventually I found a Lely spider-wheel rake like the one used by Geoff Morton. These machines were a later development than the side-delivery rakes, so I stood a better chance of finding parts, and apart from the bearings which allow the wheels to caster and the spider wheels to move up and down, the only moving parts are the wheel bearings, so there is very little to go wrong with it. With this machine, I just put it to work behind the hitch cart, with both the shafts and hitch point offset to the side. Although this is a harder pull than the tedder, a lot can be done by adjusting the spider wheels to only just touch the ground, so

Rowing up hay with the Lely rake using five spider-wheels in line.

it is the hay more than the ground that makes them rotate. The other adjustment is to set the whole rake at an acute angle, so the spider wheels only just tip the swaths over, which also minimises the tendency to twist the hay into a rope. With narrow swaths, it is also possible to use four or five spider wheels instead of six.

When turning the hay for the first time, I use two spider wheels on the nearside and three on the offside to put the two five-foot swaths closer together for tedding, and subsequently use either four or five wheels. Still, I was conscious that the horse was doing more work than necessary in pulling the hitch cart, so after a couple of years I modified the shafts from the now redundant side-delivery rake to fit the spider-wheel rake, and attached a seat, footrest and safety rail. Although the Lely company did make seats for these rakes, which, in their native Netherlands were pulled by a single horse in chain harness, I was unable to find out where the seat was attached to the machine. In the end I put it over the off-side wheel which does not carry much weight, but having done it, I am not sure whether it would have been better over the back wheel. When driving from the present position I have direct contact with the horse's mouth, but when I turn to the right and the horse gets nearer, I have to take up the slack in the reins very quickly. If the seat was at the back, this wouldn't happen, but it would need a rein guard in the middle of the machine to keep the reins out of the spider wheels. (I later discovered that the seats were originally at the back.)

Turning two swaths separately, the horse now between the shafts of the turner.

Mowing

When making hay with a single horse, turning and tedding are the easy parts. Mowing presented a greater challenge, but with little spare money and suitable equipment hard to find, in the beginning I was happy to do what I could with the horse and not worry too much about the things I couldn't do. Anyway, cutting the hay is the least important thing to do oneself, because unlike baling or turning, if you can't do it one day, the crop is not going to spoil if it rains. Having my own mower, however, would allow me more control over the process, especially as the contractors switched to larger mowers, leaving swaths too big for my horse-drawn equipment to tackle.

In my search for a solution to mowing with a single horse, I had a few options. The first was to find an old mower for a single horse, but it is rare to find a good one, and even if just one part broke, the chances of finding a replacement would be minimal. A further complication is that the single-horse mowers cut a swath of three foot or three-foot-six *(0.9 to 1m)*, which would not match the working widths of my turner or tedder. The second choice would be to find an old two-horse mower, but with any British mower the parts problem is the same as with the single horse machines. With some American-made mowers, however, the parts situation is completely different, and although any parts would need to be imported from America, that was still a possibility. Another option was to use a mower with a more modern cutter bar. Like Joe Godderidge, I could have tried using a Busatis cutter bar, but trying to scale down a home-made mower from a three-horse machine would mean it would cut a very narrow swath. A mower using the same cutter bar but driven with a small engine would have been another possibility, but although I am happy enough cutting and welding, I did not fancy the precision metalwork or the cost of an engine for an experimental mower. The last option was to buy a motorised hitch cart to use with a tractor mower, but the expense would be considerable, and although in theory I have nothing against using engines with horses, it was not something I wanted to do if I could avoid it.

In the end I decided on an old American two-horse mower. For the availability of parts, the choice was between John Deere and McCormick Deering, but John Deere mowers were never imported into Britain, so that only left McCormick Deering, who exported horse-drawn mowers until the Second World War. From 1939 the trade switched to tractors and machines to go with them, so the no 9 mowers which started rolling off the production line that same year, probably never made it across the Atlantic. This left me with just the no 7, and in due course I found one.

My original intention was to completely rebuild it so it would be in the best possible working condition, and although I was still unsure whether this would be too much for one horse, I could always make the cutter bar shorter if necessary, or even disengage the clutch mechanism and install a small engine to run a belt to the pitman wheel. Cleaning the no 7 mower revealed the frame, wheels and gearbox to be in reasonable condition, but the cutter bar needed rebuilding, and there were numerous missing or worn parts, including the draught bracket, which suggested the mower had had a hard existence being dragged behind a tractor. Although these problems weren't insurmountable, when I inquired about buying replacement parts, it came to not much less than a reconditioned mower, which would probably be a better machine than my first attempt at restoration. So, the rebuild was put on hold while I reviewed my options, the new option being to import a reconditioned mower from America.

With this new possibility, I would no longer be confined to having a no 7, and the no 9 perhaps had some advantages. Although many horse farmers in the 'States are equally happy with either model, the fact that the company had gone to the trouble of developing the no 9 suggests they could see that there were improvements to be made. Even if the changes were small, any improvement would increase my chances of success with a single horse. It also made sense that the two-stage gearbox of the no 9 would absorb less power than the three-stage gearbox of the no 7, and its position behind the axle would help in reducing the weight on the horse. Another advantage in getting a reconditioned mower was that I could specify inflatable rubber tyres rather than solid wheels. In a dry field, rubber tyres reduce the draught of a vehicle by 45%,[1]

though with a mower, there is obviously more happening than just pulling it along out of gear as if it were a cart. For example, tests using McCormick Deering no 7 mowers [2] show that with a 5 ft *(1.5m)* cutter bar, the rolling resistance (the amount of force needed to pull the mower when out of gear) is 94 lb of force, the mechanical resistance (the additional force needed to make the whole machine work) is 74 lb, and the force used in cutting the crop is 152 lb, making a total of 320 lb. *(1.4 kN)*

If, by using rubber tyres I was able to reduce the rolling resistance of my mower by 45%, from 94 lb to 52 lb, reduce the mechanical resistance to 67 lb by using a 4'6" cutter bar (the figures suggest this is mostly proportional to the cutter bar length) and the resistance in cutting the crop due to the shorter cutter bar to 139 lb, that would theoretically reduce the total force required to 258 lb, a reduction of 20%. Another way of looking at it is that using rubber tyres would have a similar effect to cutting six inches off the end of the cutter bar. Of course, this was all theoretical, but the extra cost of rubber tyres seemed well worth the possible improvement in draught.

The new mower fresh out of the crate, still without any shafts.

Despite the advantages of a reconditioned mower, ordering it from the 'States was still a shot in the dark, but after corresponding with a couple of firms, I placed an order for a number 9 with inflatable tyres and three topless guards. A year later the mower duly arrived, resplendent with new paint, parts and possibilities. Although getting an as-new mower theoretically meant it would be ready to go, it took quite some time to get it ready for work. The first task was to remove the paint from the wearing surfaces of the cutter bar and to adjust the wear plates and clips, so the knife ran freely, which makes a big difference to the draught. A job I had not expected to do was cleaning the gearbox, which judging by the colour of the oil was not a job included in the restoration. My concern that this had not been done was allayed when I cleaned the gearbox with diesel and a toothbrush, revealing the gears to be in good condition.

The other job was to fit a pair of shafts, either directly onto a stub pole, or in front of a tongue truck or dolly wheel. Originally these mowers were designed with a pole, the transferred weight of the front of the mower being carried by the neck yoke, and thence to the horses' necks. Although this weight can be minimised by having the neck yoke high enough and lowering the point where the cobbletree attaches to the mower,[3] having a dolly wheel or tongue truck at the front of the mower means the horses' necks are only carrying the weight of the pole. The down-side of the tongue truck is that the extra wheel(s) cause extra draught. If the combined weight of the wheels, the bracket, half the weight of the pole, and most of the weight of the cutter bar amounted to 150 lbs, *(68 kg)* even with a rubber wheel, this could equate to about another 9.5 lbs *(42 N)* in draught. For a pair of horses this may be a price worth paying, but for a single horse I would do well to save this extra load.

The next decision was where to fix the shafts, either in line with the pole or offset to the side. The original position of the pole, it seemed to me, was mostly determined by the need to have enough space for the offside horse to walk between the standing grass and the pole, rather than being placed at the centre of draught, nearer the grass, which is where the shafts of single horse mowers are attached. In practice, the position of the pole in the mower's frame limited any lateral adjustment that could easily be made, so I built an additional bracket mounted immediately to the side of the stub pole. This moves the draught point four inches *(10 cm)* nearer the standing grass. With the mower in use, the shafts remain central on the horse and do not rub either side, suggesting that the line of draught is good. The final addition was a wooden support to hold the shafts up when not in use, so I do not have to lift the heavy shafts when yoking up.

The shafts and draught bracket mounted to the side of the pole.

The mower was now ready to go. Even though my horse is used to a variety of sounds and sensations, whilst the mower was up on axle stands and I had Molly in the vicinity, I took the opportunity to turn the wheels so she could see and hear everything move and hear the clicking of the ratchets in the wheels when turned backwards. Even so, the first time I put the mower in gear, she stopped after two steps, probably because of the unusual vibrations. So, I quickly said "whoa" and relaxed into staying still for a moment or two. Then with an upbeat, "go on, me lass," followed after one step by a more commanding but encouraging, "go on," we went and cut hay.

The amount of hay we cut, however, was tiny, because within the first hundred yards the machine must have jammed ten times. With all these enforced stops I started to wonder whether I was ever going to get the hay cut, but as we continued mowing the job became easier. I learned to push her on when I got to a spot where the grass was thicker, so the stops became fewer, but we still had to stop two or three times every round of the field. With two horses and their greater reserves of power, some of these stops would not have happened, but cutting hay that first year was hard work.

Once that year's hay making was done, I decided to try and improve the mower's performance. Like many American horse farmers, I had three topless guards at the near end of the cutter bar. The usual reason for using topless guards is to avoid the need for a grassboard, so the cut hay lies the full width of the swath, exposing the greatest surface area to dry. Without a grassboard to push the cut hay away from the standing grass, should any cut hay get in the cutter bar on the next round, the topless guards chomp through it rather than jamming the machine. My reason for having the topless guards was slightly different, as I do use a grass board, but if I wanted to reduce the required power a little by driving slightly away from the grass, I could

Above - The cutter bar in its original configuration with three topless guards. Left- the modified cutter bar with topless guards along the whole length.

cut six inches less, and if I did drive a little inaccurately and cut into the swath, the knife still wouldn't jam. On my many stops to clear the cutter bar, I noticed that the knife never clogged where the topless guards were, unlike on the conventional fingers where a mouse nest or other mess of grass can get stuck between the points, preventing the knife from reaching and cutting the standing grass. So why not put topless guards along the whole length of the cutter bar? Although I had never come across anyone doing this before, it seemed worth a try, so I ordered a complete set. This makes the mower look a little different, but the first outing proved the value of the change, forced stops being much less frequent.

Mowing is now a much more pleasant job, though it is still hard work for the horse. Often, I cut a few times around the outside of a field in the evening, so the horse has a long rest before cutting the rest the following morning. Even if there is a possibility of some rain overnight, as long as the forecast gives a good spell of weather to follow, I will still cut some the night before, because by the time the remainder of the hay is cut, the bit that may have been rained on is as dry as the hay in the centre of the field. I am also very careful with Molly's shoulders and usually wash them with salty water after this hot and heavy job, but the most important thing is to have the horse fit before starting to cut the hay.

Drying the Hay

For my neighbours, my hay making activities must seem peculiar, because besides using a horse, at any time I might have loose hay in a stack and some being baled, some drying on the ground, some on tripods and occasionally some in cocks. The reason for using these different methods is because I sometimes need to be away from home during the hay making season, and especially in the unsettled weather of the last few years, these approaches allow me to make hay when the weather is not ideal.

In the beginning, however, I made hay by the usual method, turning it once a day with the horse before getting it baled. When making hay on the ground, there is a divergence of opinion about when it should be turned. Some like to leave it untouched for a day or two after mowing, saying that it is damaged less by rain if it has not been moved, but others prefer to move the hay immediately and often, to get it dry as quickly as possible, and there have been studies [4] which show that the best hay is produced by making it quickly. Even if the hay does get rained on, having it drier before the rain comes (because it has been turned) still means that it is fit to carry sooner and is therefore of better quality.

When I started out, I usually left the hay lying in the swath until the day after cutting, when I turned it with the side delivery rake to expose the damp bottom of the swaths to the air. Once I had a tedder, I then went through the hay straight after turning, as this is the time when the leaf is least damaged by being tedded. On subsequent days, turning it once a day was often enough to get it sufficiently dry to bale by the fourth day. In recent years, however, with the catchy weather, I turn the hay soon after cutting to expose the damp underside and then ted it straight afterwards to separate the stalks. It is turned and tedded again on day two, then on the third day it might only need to be turned, and sometimes it can be carried later the same day.

Tripods

In parallel with making hay on the ground, nearly every year I make some hay on tripods. The attraction of this method is that it only needs one day of reasonably good weather to dry the grass enough to put it onto the tripods, and then the hay takes very little harm, usually coming out green, dry and smelling beautifully of hay two weeks later. The disadvantages are the initial effort in making the tripods and the time it takes to load the hay onto the tripods.

For a period in the 1950s, tripods became quite popular, largely due to the efforts of the Scottish agriculturalist, Alexander Proctor. I have talked to two people who used tripods, way back in their youth. The first told me that when Mr Proctor visited and showed them how to load the tripods, they had marvellous hay, but when they tried it themselves the following year, the hay was no good. My other informant, Ron Creasey,[5] worked in East Yorkshire where they used tripods for hay and for peas, but these 'tripods' actually had four legs and did not have any air vents at the corners, and in his experience the hay was not put onto the tripods until about half dry. My education in haymaking with tripods has therefore been by trial and error, with snippets of information gleaned from books [6] and a study done in the 1950s by the National Institute of Agricultural Engineering. [7]

There are two essential requirements for making hay on tripods; the tripods must be built properly and the grass must be dry enough when it is put on the tripod. If the grass is cut by mid-morning and tedded immediately afterwards, on a hot sunny day or a warm, breezy day, the grass will be dry enough for the tripods to be started by late afternoon. However, if the grass was rained on the night before or if the day is cloudy without much wind, the hay may not be dry enough to load onto the tripods until the following day. The study by the National Institute of Agricultural Engineering showed that the moisture content needs to be down to 55%, though they suspected that in skilled hands 60% moisture would not be too much. My experience is that the grass should not feel damp on the surface, though if there are a few damp lumps, they can be put on the very top of the tripod where they will still dry, and in the meantime, they help weigh down the top and help it to settle.

The procedure I follow is to turn the hay straight after cutting, keeping the swaths separate, then put the horse in the tedder to fluff it up. It is then time for the horse to eat and rest until

Building Tripods

A good size for a tripod is to have the legs 7 feet long. *(2.1m)* Using round saplings rather than square sawn timber is preferable as it is stronger for its weight, so it can be thinner and lighter. The top of each pole is drilled about an inch and a half *(4 cm)* from the end and a loop of 1/8" *(3mm)* wire pushed through and the ends twisted together. The wire triangles which carry the hay are also 1/8" *(3mm)* thick, the lower wires being about 40 inches *(1m)* on each side, the top wires 21 inches *(53cm)*. I find I need about a dozen tripods for an acre *(28/ha)*.

1. Place the tripod on flat ground with the legs about 4 feet *(1.2m)* apart. Lower the first wire loop into position. Adjust the position of the poles until the wire is between 18 inches and 2 feet *(45 to 60cm)* from the ground, so the hay will scarcely touch the ground.

2. Lean two plywood sheets about 16" *(40cm)* square on both sides of each leg to form the air vents.

3. Start by placing handfuls of hay on the wire each side of the vents. This helps to keep the wire in place. I always bend the inside part of each handful of hay back out again to help keep the air vent open. Continue putting the first layer of hay round the rest of the tripod with your hands. If the weight of hay allows the hay to reach the ground, lift the wire and move the bottom of the poles outwards.

4. Once the first layer is on the wire, switch to using a fork and gently put small forkfuls on top of the first layer. Load the wires evenly and avoid putting too much on at once. Try and keep the outside nearly vertical.

5. To help keep the sides upright, after putting on the second layer, put a forkful round the outside of each pole, so it only rests on hay and not the wire.

6. Keep going round the tripod adding small forkfuls, putting some on the outside of each post when necessary. When the hay is about three foot high, lean in and check there is still a good open air space in the middle.

7. When about chest height, put on the next loop of wire. This should just gently press down on the last circle of hay. Put your arm down the middle to check the air space is still open.

8. Larger forkfuls can now be put on the tripod, forming an even circle round the outside of the poles, leaving the centre only filled with air. It is at this height that the tripod's capacity is at its greatest, as long as the walls have been kept fairly upright lower down. If you have a helper carrying hay to you, this is time for them to remove the plywood formers and set up a new tripod, and put on the first layer with their hands.

9. Continue adding forkfuls of hay until you get level with the top of the poles. It is tempting to put some hay over the top while the poles are still sticking up above the hay, but doing this reduces the amount of hay on the tripod.

10. Once the top of the poles is reached, two or three wide, well-made, flat forkfuls of hay are carefully placed centrally over the top of the tripod.

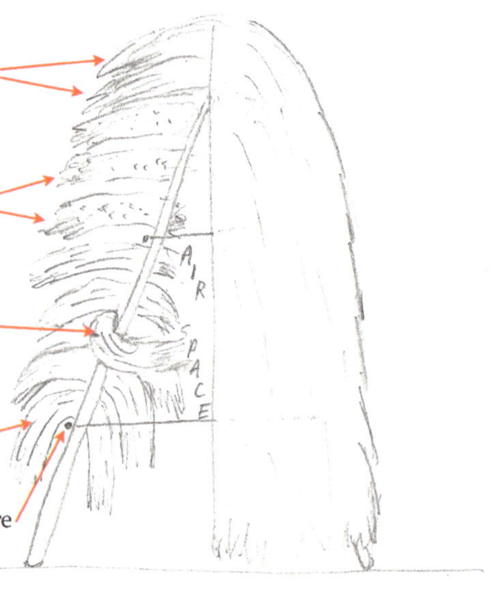

mid-afternoon when I put her in the turner to bring the hay together ready for loading onto the tripods. First, I turn a clear strip where the tripods will be set up, and then turn the hay on either side towards this gap, leaving the windrows close together but not touching so they can continue to dry. Depending on the thickness of the crop, I rake three to five windrows (six to ten mower swaths) to make a row of tripods. Using a trip rake or buck rake to bring the hay to the tripods would be appropriate if working in a gang of four or five people, but I usually work on my own or with one other person, so it is easier to do all the horse work before starting to build the tripods.

Hay dried on tripods takes longer to dry than hay made on the ground; even in hot sunny weather a week would probably be the very shortest time, but two weeks is normal. Although the outside gets bleached by sun and rain, 95% of the hay remains green. Once the hay is dry and the top of the tripod is pushed up with a fork and put on a vehicle, the rest of the hay is easy to fork in good sized forkfuls from each side of the tripod. The only slightly awkward bit is getting the bottom layer off the wire.

There are some variables in tripod making. Following Ron Creasey's experience, I have also tried using four poles. These quad-pods work alright and carry more hay because there are four sides instead of three. It is also easier to keep the air gap in the centre, but with the extra length of wire and the extra weight of hay, the bottom also tends to sink towards the ground. The four-poled structures are also harder to put up if the ground is uneven, and it is more difficult to keep the bottom wire in place as you start loading it with grass. Also, when taking the poles and wires to the field when all the wires are together in a bunch, the four-sided wires are harder to separate as they tend to bend back on themselves, whereas the three-sided wires stay flat. Because of these disadvantages, after a couple of years I remade my quad-pods into tripods. Another variable is the length of the poles themselves. I have used poles from five to eight feet long *(1.5 to 2.4m)*, but always with wires of the same length. This has made the tripods into more or less dumpy shapes. Using poles less than six foot *(1.8m)* long means the amount of hay per tripod is reduced, but if you only have five-foot poles and not much hay to make, they are fine. When the poles get to eight foot *(2.4m)*, it is harder to put those small forkfuls around the outside of the poles, because they are at a steeper angle, so the tripod tends to get narrow too quickly, so reducing its capacity.

Hay Cocks

In comparison with making tripods, putting hay into cocks is easy. The important thing is to use the effect of gravity which has made the hay settle relatively flat in the windrow. Then as you put your fork into the hay, lift it upwards and sideways to separate it gently from the hay next to it. This forkful can then be put on top of the next part of the row, and both layers lifted together in the same fashion. This larger forkful is placed flat on the ground to start the cock, and the process is repeated, placing flat forkfuls centrally on the partly built cock until the cock is high enough. Especially when making a big cock, the forkfuls need to be wide enough so that each addition forkful of hay fills the entire circle. If you use smaller forkfuls, the cock becomes unstable. Giving the finished cock a tap on the top probably helps it settle.

There are two situations where I still make hay cocks. If it starts to rain when making tripods, I change to making cocks, as any external moisture would lead to moulding in the tripod. The other time I cock some hay is when I have some drying on the ground and it gives rain overnight, especially if the hay is already nearly dry. In this case it can finish drying in the cock before being taken to the stack, but if the hay has only been cut for a day or two, the cocks will need to be taken apart the next morning and spread on the ground to dry. Although this is a lot of work, the quality of the hay is preserved because only a small percentage of the hay in the cock gets wet from the rain. If a cock of nearly dry hay gets rained on, I take the wet top off and spread it out to dry, and regardless of the weather I usually turn the whole cock over so the damp absorbed from the ground can dissipate before loading onto a vehicle. To make hay really quickly, in terms of days but not in hours of work, putting hay into cocks is probably the quickest method, as the hay dries quickest when spread out during the day, but only a small part of it gets damp with dew whilst in the cocks overnight.

Norwegian Drying Fences (Hesjer)

In Norway, most hay used to be dried on drying fences, a method which allows hay to dry even in wet conditions. Once the hay is cut, a row of posts are driven into the ground at about six foot *(1.8m)* intervals, with a stout peg and a guy rope at each end. The first wire is strung along between the posts, about a foot from the ground, the wire being wrapped one turn round each post. Subsequent wires are about 9 inches *(23cm)* apart.

As each armful of hay is gathered, it is first thrown a foot in the air, then picked up again loosely before being placed on the wire, starting at the bottom.

The distance between the wires allows plenty of air to get through the hay.

As the hay settles, the top layer forms a roof, protecting the lower levels from the rain. Because the hay is so loose, it dries from the effect of air movement rather than the sun.

As I finished writing this, we had experienced the wettest April, May and June ever recorded and the wettest summer (June to August) in a hundred years. Some hay which I put on tripods in mid-June was rained on every day except one for a month, often with over an inch falling in a day. On the 12th of July I threw the hay on the ground for the sun to finally dry it enough to put in a stack, but the bottom half of each tripod had gone mouldy and was unusable. Since the bottoms of some of the tripods were actually standing in water and there had been no warm air blowing through them, this was not surprising, but it shows that no method is perfect.

After yet another week of rain, we had another dry day, so as an experiment, I cut half an acre on wet ground, and to avoid getting the hay muddy, I carried it with a fork and put it on a Norwegian-style fence whilst still wet. After a damp week the hay was dry enough to stack, but a day later dry weather came at last, so the hay on the fence had to stay there for a few more days so I could make the rest of the hay in the usual manner.*

Bringing in the Hay

Over the last few years of hay making, the mowing, turning and tripod building has settled into a fairly comfortable pattern, but the process of getting it all together for the winter is still developing. In the beginning I did what everyone else does and got it baled, sometimes by a contractor or a neighbour, and sometimes I borrowed a tractor and baler. In some ways I liked having someone else to do this job because it took the pressure off doing it myself, especially when worried about someone else's baler breaking down. With hay made on tripods, there was no worry because the hay would not spoil on the tripod, and I could often bale when no-one else was baling, driving the baler from tripod to tripod and forking the hay in front of the pick-up reel.

Bales do have some obvious advantages, being easy to handle, load and stack, but they also have disadvantages. Besides the cost of baling and not being able to guarantee access to a baler when I wanted it, the Dutch barn I rented was over a mile away, so clearing the field with a horse was slow. For a few years I thought about stacking the hay loose, but in the midst of haymaking when all the hay was on the ground, I chose the familiar solution, continuing to put it into bales. My reluctance was mostly fear of the unknown, having never had the opportunity to be present when a haystack was being built, but one year I decided to try one small stack, following the instructions in Lynn Miller's book 'Haying with Horses.' [8]

The success of this first stack encouraged me to do more, and now most of my hay is stacked loose. My stacks are usually quite small as I rarely have more than an acre or two ready at the same time. Small stacks have some advantages when the hay is forked onto the stack with

* *I have successfully made hay on drying fences several times since then.*

Building Haystacks

Traditionally, haystacks were built on a layer of brushwood to prevent moisture from rising and spoiling the hay, but pallets and other lengths of wood work just as well, allowing air to pass under the stack. When building the stack, each layer, or course of hay needs to be higher in the middle than at the edge. For the first course, put a good big forkful of hay in the middle, then place subsequent forkfuls in a circle around it as you walk round, overlapping the previous hay and keeping the centre high.

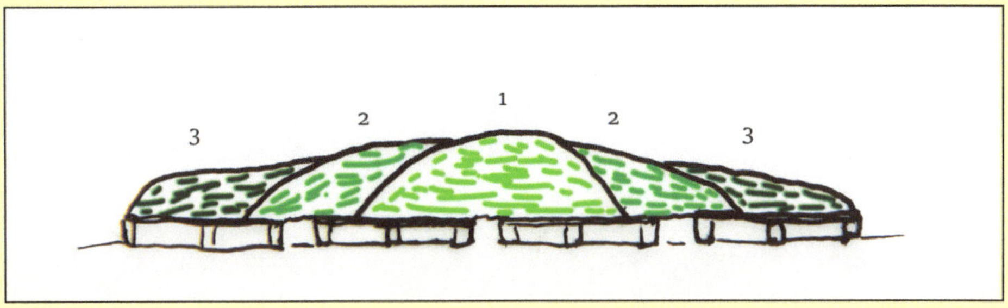

Even a very small stack will require two circles of hay. At the edge, make sure the outside forms a good round circle. At this stage, the person on the load needs to curtail their enthusiasm to fork the hay down quickly, so the person stacking has time to make the base of the stack solid and round.

For subsequent courses, start at the outside. Put the hay down in big forkfuls and then stand on the lump you have just stacked. This compresses the hay, and you can feel how solid it is under your feet. If it feels a bit slack, step back and add another forkful of hay. If someone young or inexperienced is forking small forkfuls, put a few together before putting them in place and stepping forward. It is essential to keep a good solid wall of hay round the edge. Once the outside circle is complete, the next circle of hay slightly overlaps the outside circle. This helps to trap the outer circle of hay in place and allows any water to run away towards the outside. Keeping going round and round until you reach the centre, always keeping the centre higher that the outside.

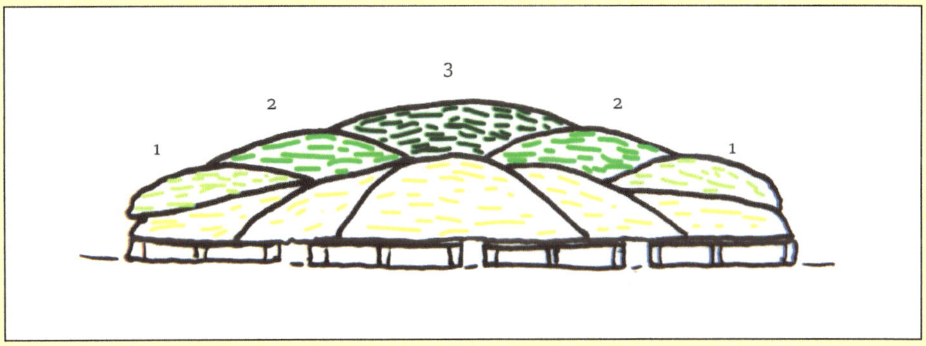

Having done one course going around clockwise, stack the next one anti-clockwise to help the stack hold together, alternating each course. If you are getting loads from a mechanical stacker which land on the middle of the stack, the centre probably gets compressed enough and looks after itself, but if someone is forking hay to you, you need to walk round the middle too, to make sure the hay will not sink lower than the outside as the hay settles. If you find there is a slack place in the build, pop an extra forkful or two in the hole, and if the centre is starting to look too low in relation to the outside, put some more hay in the middle before starting the next full course.

When stacking around the outside, it is easy to bring the sides in rather than keeping them vertical, especially on the side furthest from the load. On all except the smallest stacks, it is an advantage to have someone else on the stack to fork the hay beside you, so you only need to rotate each forkful a quarter turn round your body, allowing you to concentrate on stacking rather than reaching for the hay. When you get down off the stack to get another load, walk round the stack to see how even and round it is. If it sticks out beyond the circle in one place, raking a bit out of the sides with a fork can help, but if there is not enough anywhere, you can't add hay to gain width.

Once you are ready to draw in the sides to form the roof, keeping an even circle is more important, because it is at this point where any slackness in the build will allow water to penetrate. Avoid bringing in the sides too quickly, because as the hay settles, the extra weight in the middle tends to push the centre down more, so the top becomes flatter over time. This is why, in each course, the centre should now be kept even higher relative to the edge. Continuing to work in circles, your helper will eventually not have enough room to stand, so should get down off the stack. As the top is reached, only three or four big forkfuls will complete the course, until eventually, one big well-formed forkful completes the top.

Once the stack is finished, rake down the sides with a fork to tidy any loose bits, then throw two ropes over the top and tie a brick at each end. This prevents the top getting blown off in the first few weeks until the hay settles. Once the top has knitted together, if the stack has been built correctly, the rain is shed from the top with only a minimum of wastage.

A small haystack in mid-winter, half having been cut away with a hay knife. One of my first stacks, I brought the sides in too quickly, reducing the capacity of the stack, but it was nonetheless very weather-tight.

pitch forks, especially when relying on children or other inexperienced people, but the smaller stack does have a greater surface area exposed to the weather compared with a larger stack. Another disadvantage of very small stacks is that they can easily become unstable, ten feet *(3m)* probably being the absolute minimum diameter; otherwise, when the stack gets up a bit it is like standing on a jelly, and I have had the top of a small stack slide off, taking me with it. This is most disheartening, not to say potentially dangerous, and a good reason never to leave forks, ladders, bricks etc. around the bottom of a stack. If I was going to make a stack smaller than ten feet *(3m)* in diameter, I would now build it around a post. Because much of my hay is fed outside, I place the stacks near the track that the horses use in winter, near enough so I don't have to carry it far, but far enough so they can't reach it, avoiding trees and any area where water gathers.

For me, the time spent stacking loose hay in the field is saved by not having to take bales a mile away to the barn, and having the stack in the field where the hay will be fed saves more time in winter. Gathering loose hay with a single horse, however, is significantly slower than collecting bales. When I make hay on tripods, getting the hay onto a vehicle is straightforward as the hay has settled on the tripods so is easy to fork off in good sized forkfuls. With cocks, the hay has also settled so is easy to fork onto a load, but both methods require an additional stage of hand work compared with gathering hay from the windrow. To speed up loading I have used a hay loader, but found that although one horse can pull it, it is too much to pull slowly enough for the comfort of man and horse. I do sometimes use it behind a borrowed tractor, with hay ladders at each end of the vehicle, and because the rulley is narrow, three ropes tied end to end along on each side to stop hay falling off as it falls from the hay loader.

Still on the search for a better way to gather hay with one horse, after reading an article about a homemade buckrake for a single horse, I decided to try making one myself. Unlike the two-horse buckrakes where a horse works on either side of the collecting basket, or with some American models they walk behind the basket, in this design the horse walked beside the hay swaths with the basket behind the horse and offset to one side. The machine in the article required the operator to move backwards along a plank to lift the tines clear of the ground, but having studied the different American buckrakes where the pull from the horses lifts the basket, I wondered whether I could incorporate a lifting mechanism on a single horse machine. In the end I used two features from some American horse-drawn big bale movers: a braked axle and a sliding drawbar to lift the basket, in conjuction with my hitch cart. Although attaching shafts directly to a buckrake would have made the whole set up shorter, using the hitch cart allows the buckrake drawbar to be extended when the brakes are applied, turning the horse and hitch cart's forward motion into a lifting action, whilst any side draught caused by the buckrake being off-centre is countered by the wheels of the hitch cart. As when tedding hay, the hitch cart shafts are offset to one side, so the horse and hitch cart are not running over any hay and the drawbar of the buckrake is directly behind the horse.

With the tines of the buckrake lowered to gather hay, the drawbar, which consists of a square tube inside another square tube, is in its shortest position, held in place by this spring-loaded catch. Once the basket is full, I stop the horse and pull a lever to lift the catch and put my foot on the brake pedal to keep the buckrake still. When the horse moves forward, the inside tube of the drawbar slides forward, and as it does so, it connects with a lever at the back of the machine which lifts the basket.

The rear of the buckrake showing the long diagonal lever, which when pulled forward by the drawbar, pivots at its centre and pulls the top of the basket backwards, so lifting the tines. Once the rear of the drawbar reaches the end of its travel, another spring-loaded catch holds it in place, keeping the basket in the raised position. As I hear this catch click into position, I take my foot off the brake pedal and we head for the stack. To lower the basket, I stop the horse, put my left foot on the brake pedal, my right foot on the pedal to release the rear catch and back the horse a step. The drawbar then slides back, the tines are lowered to the ground and the front catch clicks into position again. Continuing to back up with the brake off, the hay slides off the basket.

Of all my attempts at metal fabrication, this mechanism probably gives me the greatest satisfaction, but there are some downsides of the buckrake. Firstly, taking relatively small amounts of hay to the stack it is not very quick. There is also a lot of backing and messing about for the horse when gathering hay, because after a while the hay starts to ball up and roll under the tines, so you need to back up and gather some more and have this happen again before going to pick up both lumps and taking them to the stack. Also, on uneven ground the tines can dig in, though up to now I have been quick enough to stop before a tine has broken. Once I get to the stack, the hay is obviously left on the ground, so without any mechanical stacker it all needs to be forked up to the height of the stack, whereas on a vehicle it is already part of the way up.

The gathering of hay at the moment is the weak part of my system, and most years I use a mixture of methods. When I use tripods, forking the hay onto the vehicle and stacking it loose is the obvious next step, but when the hay is in the windrow the rest of the family prefer bales because the gathering is quicker than borrowing a tractor and using the hay loader. Overall, however, because the hayloader is seven feet wide *(2.1m)* at the bottom, it can pick up the hay from two swaths without having to bring them together into a windrow first, so the hay is gathered in one operation rather than three when rowing up, baling, then gathering the bales.

Postscript

Now, in 2025, I still use the same implements as I did a dozen years ago, but for eight years I had two horses at work. For turning and tedding I could have continued to use one of the horses on its own, but by using the hitch cart I could use both horses, a good training schedule for the younger horse to hear all the rattling of the tedder and to see the turner near at hand when doing tight turns, all with the support and calm attitude of the older horse. Besides that, using two horses made the work easy so resting times, even on hot days, could be kept to a minimum. By comparison, putting two horses in what was designed as a two-horse mower made life very much easier, turning a hard pull for a single into a moderate pull for a pair, making a huge difference to the necessary resting times.

Nowadays, old hay making machinery in working condition is even harder to find than it was when I started twenty-five years ago. Especially considering the number of repairs I needed to do and the time spent hunting for spare parts, I sometimes wonder what I would do if was starting again, the best, though initially expensive solution being to buy new equipment. Compared with using two or more horses, the draught of any implement is more critical when using a single horse, so most tractor machinery is out of the question. However, there is some equipment well suited for use with a single horse. For light draught, the new Italian side delivery rakes [9] are exemplary, though for British conditions they would need some modification to allow the middle of each raker bar to be removed for turning two mower swaths independently. Unlike the spider-wheel turner, they don't bunch up the hay when turning at the ends of the rows, nor do they crab downhill on side slopes. There are also modern versions of the reel-type tedder, made in

Above - A Repossi side-delivery rake, demonstrated at the Pferdestark event in Germany.

Below - A fork-type hay tedder made by Hoover Hay Tedders. This extra-wide version with a width of 5.4 metres can be folded to get through gateways, but is nonetheless pulled with ease by two Fjord ponies.

America, and an impressive fork-type tedder which is exceedingly light of draught. For mowing, the major advance has been the use of the German-made ESM cutter bar which has two knives working in opposition to each other, which minimise vibration, and because each knife section is spaced irregularly along the cutter bar, different parts of the knife are at different stages of the cut, reducing the power required by a third, a huge benefit when using horses.

Above - Christoph Schmitz's mower incorporating an ESM cutter bar.

Below - I and J's traditionally-styled mower. Because the 2.4m ESM cutter bar is heavier than a traditional one, the mower is equipped with a dolly wheel to take weight away off the pole. This same mower, but with a shorter cutter bar can be used with a single horse.

Opposite page - Dairy Shorthorn cows grazing at Botton village.

Chapter Four
Comparative Craftsmanship

Small Farmer's Journal, 2007

Around the time I was getting to grips with using old machinery on my own small place, many full-time farmers were facing tough times. For horse farmers, time and financial concerns only intensified the pressure to use a tractor for some work, or a hitch cart with an engine, and prompted a discussion in the Small Farmer's Journal about their threat to rural craftsmanship.

When we are working on our farms and smallholdings, it is easy to feel that we are toiling away in isolation, sometimes struggling as we try to find time to fix the fence or sort out that recalcitrant machine, in between doing all the regular chores. The welcome arrival every few months of the big brown envelope containing the Small Farmer's Journal, in this house and I am sure in many others, is the cue to put on the kettle and flick through a few pictures, read a letter or two, and, if other jobs are not too pressing, delve into one of those detailed articles which put the Small Farmer's Journal into a class of its own.

Besides being a source of information, reading the Journal also helps us take a step back from our immediate concerns and makes us realise that we are connected by slender threads to other people who are doing similar things in far-off places. One of those connections which I hadn't really thought about, raised in the editor's discussion about horse-drawn machinery and the place of craftsmanship, was brought into sharp focus when he compared the situation with that of a violin maker. The reason it struck a chord is that when I am not working my Percheron mare or digging out ditches, I spend much of my time making violins. The two activities, doing farm work and making violins, may seem to have very little in common; indeed, I am sometimes asked how I can do both the fine work of instrument making and the rough physical work of farming. But for me it is all part of everyday life and although the differences are most obvious, in terms of craft, the similarities are more interesting.

When I started out as a violin maker, I made instruments using nothing but hand tools. In the beginning there wasn't any spare money to buy expensive equipment, but that didn't matter as the hand tools were perfectly adequate for the job. Then one day I bought a bandsaw, not a small one that would do most things, but one big enough for the odd occasion when I needed to cut a big piece of wood to make a cello back. Once I had learned how to use it, the bandsaw did save some time and effort but compared with the time I spend at the workbench, not all that much. The rest of the time I continue to use hand tools; it is still fundamental to what I do. Having the right tools and knowing how to use them so that it feels instinctive allows me to concentrate on what I want to make, the tools disappearing from conscious thought as I shape the wood. Using simple tools is part of what makes it a craft.

The arrival of the bandsaw, though a visual intrusion into the workshop, didn't mean that I was no longer a craftsman, nor does it diminish the quality of my instruments. The way I use my bandsaw is comparable to the hired tractor and baler which a horse farmer might have come to bale his hay after cutting and turning it with horses, a quick solution for one job while the rest of the time everything else is done with the horses. But there is an important difference, because the bandsaw belongs to me; it is my own tractor; I don't have to arrange for it to come and it doesn't go away when I've finished.

Although the bandsaw is useful, its saving grace is that it is not very useful, because it only does one thing, it cuts; it does not plane and it can't apply varnish, so on a finished instrument it is impossible to tell whether it was used. More importantly, it doesn't change the fundamental nature of how I think about or do my work.

There are some instrument makers, however, who make greater use of machines, little routers to make the groove for the purfling and profile machines to carve the basic shape of back and belly. For me, this is a step too far, it removes too much of the connection to the raw material and each instrument is likely to be much like the last one, because once a machine is set up, it is much easier to use it again on the same setting, even though the craftsman's instinct is to subtly change things from one piece of work to the next. This is the nature of a craft, the expression of the craftsman's intention in the work, which develops and changes day by day.

Just as using hand tools is fundamental to my work as a violin maker, I suspect most of us who use horses do so, not just because we like them, but because we appreciate the close connection with our work. We like that our tools are simple and easily repaired, that we can see and feel what's happening between the soil, the horses and the implement. We like the fact that to work effectively we need to understand and balance all the variables, the harness, the hitch, the implement adjustments, the tension in the lines, the different characters of the horses and our own preferences. That is the nature of a craft.

The big difference between violin making, perhaps the epitome of a craft activity, and farming, is that most jobs on the farm can be done with big machines. So, unless a horse farmer is particularly innovative, skilled and single-minded, he will be drawn or pushed towards using powered machinery to do at least some of the work.

If a horse farmer does yield to the pressure and uses a tractor or a hitch cart with a mounted engine, the question, and perhaps some of the answer is, how much? If a tractor gets to be used an ever-increasing amount, as happened in the 1940s, the horses won't be fit enough to do the work we know they could do, and the less they are used, the less useful they become. With a power cart, the change will not be as great, but it may seem like a renunciation of the ideal of using quiet and sustainable animal power. If used only one day in twenty, it might, like my bandsaw, only make a small change to the way we farm, but if it is in near daily use, not only will we lose a measure of independence but also some of the intimate connection with our work. Depending on our own circumstances, that could be a price worth paying if it allows us to continue farming with horses. In the end, if the choice were to come down to using a motorised hitch cart or get a job in a factory or office, the power cart could be seen as the saviour of the farm; far better to accept a small part of the industrial process on the farm than for the world to lose a good conscientious farmer.

Deciding on an appropriate level of technology is an age-old problem, but it doesn't get any easier to solve. Perhaps that is why most of society completely ignores the question, indiscriminately embracing everything that is new. Fortunately for horse farmers, the 'new' now includes many pieces of equipment designed to be used with horses, some requiring an engine on a hitch cart or the implement itself, but many others using ground drive.

A feature of the best tools, the old faithful equipment as well as the best of the new, is their elegant simplicity. As useful lumps of metal they might not be beautiful in a classical sense, but are often beautiful in their function, allowing small forces to be gathered to greater effect, using long levers to lift tines out of the ground, wheels to turn cogs, or the horses' forward motion to create a lifting action. Implements like these embody the essence of appropriate scale and level of technology, because, whilst they must have needed plenty of ingenuity and experimentation to produce, they are easy to use, easy to repair and easily comprehended by the user. This level of understanding and involvement is important if we want to maintain control over our craft.

White Horse Machine's hydraulic hitch cart, which uses an innovative method to accumulate small amounts of energy as the horses are moving forward to lift an implement even when stationary. As the wheels of the hitch cart turn, they drive a hydraulic pump which pumps oil into a cylinder, compressing the enclosed air and thus storing the energy. When needed to extend the hydraulic rams, in this case on a two-furrow plough, a lever is pressed, and the air pressure lifts the plough out of the ground.

Even if we recognise the aspects of craft in our own lives, when we are busy, it is easy to forget the value of what we are doing; as we concentrate on the jobs which are most pressing, we tend to focus on the pitfalls and problems and overlook the special nature of our activities. Meanwhile, there are others who are fascinated and intrigued by what we do, marvelling at how a potter throws a pot or how a ploughman draws a furrow. Between craftspeople, the attraction of craftsmanship is the respect and understanding that one has for another's skill, but for someone outside the circle, the appreciation may be sheer wonder at the magical world of the craftsman. Even though it may be strange and unfamiliar for many, everyone instinctively seems to know that making things is a fundamental human activity. Despite the industrial economy only valuing anything, including craft, as whatever you have to pay for it, there is still a fundamental understanding that sees beyond this, recognising the skill, quality and the human value of craft work.

I expect most of us have experienced times when strangers have stood entranced as we work our horses, their eyes wide open as they try to take it in, just as some people are fascinated when seeing my workshop for the first time. For me, my workshop, like your farm, is normality, but from going to a harness maker's or jeweller's workshop, I know that it is not just the strange array of tools and the different smells that are exciting, but also the accumulated evidence of a job done well, carefully, skilfully over the years.

All of us can probably think of a place where small actions over time have resulted in a place of beauty; a garden where care has been lavished on the plants, a farm where skilled work and attention to detail have produced a dynamic, beautiful landscape. If this is achievable by one or two people, how much more potent it is when this care is expressed on a wider scale. In the western world this is a rarity, as I have only ever seen two examples of such a vibrant and pleasing rural community. The first was an Amish community in America, where, for the first time in my life, I experienced the quiet exhilaration of seeing an agricultural community which seemed to have a sense of balance, a human pace, as if it knew what it was doing.

A vegetable field at Botton village.

The second example was the Camphill Village at Botton in North Yorkshire, a community catering for adults with special needs. Situated at the end of a valley in the North Yorkshire Moors, the community grew over forty years to encompass a handful of farms, growing food for both the community and to sell, whilst providing work for the villagers. In amongst the farms, there were workshops making wooden toys, dolls and woven items, a glass engraving studio, as well as a bookshop, bakery and dairy. The valley was full of people, the normal means of moving about was on foot, on paths between trees, besides streams, past gardens and through fields with grazing cattle or growing vegetables. The houses built to accommodate the increased population nestled into the side of the hills, attention being paid to the quality of design as well as how they moulded into the landscape, while the farms, which provided milk for the excellent cheese made in the dairy, were run biodynamically, the hay was made loose for its quality, the small fields forming a patchwork of different crops, separated by well-maintained fences, hedges and dry-stone walls. What this brief description cannot start to convey is the beauty of this part of the valley or the feeling that you had arrived at a place which knew why it was there.

Communities like these, where craftsmanship is part of everyday life, offer a glimpse of what is possible and hope for the future. Because besides producing quality products, communities based on craftsmanship foster cooperation between neighbours and involvement in the community. When we work close to home and can buy what we need locally, the community grows stronger; when we buy vegetables from a farmer's road-side stand on the way to getting a hand-thrown milk jug from a potter, instead of getting poor imitations at a supermarket, we are not just helping the vegetable farmer and potter, but in supporting our neighbours we are also helping ourselves.

Just as a craftsman takes care of his tools, we also need to remember to value the craft in our own lives and encourage craftsmanship wherever we can, whether it is teaching someone to drive horses, play a musical instrument or bake a loaf of bread. At times this can be a hard road to travel, but if we look at the mess we have made of the world with our industrially-produced, energy-consuming lifestyles, it is clear that we need craft now more than ever. But just at the time when the world needs it most, might also be the time when preserving craftsmanship at home is most difficult; having a good feeling about using a hand tool doesn't necessarily pay the bills. But before discarding craftsmanship in favour of an industrial solution, we need to be very sure that our lack of time is not just a lack of organisation, a lack of money is not due to a lack of effective contact with people who value what we produce; easy to say when you are not chasing your own tail just to get through the jobs which must be done! If we feel we have no choice but to give up some part of craftsmanship, buying a 'bandsaw' might be just enough to ease the burden whilst keeping the core of our activity intact, but if we discard or postpone too much of what we cherish and delight in, it is too easy to be drawn into a situation where our life, our work, is only a poor approximation of what we want it to be, and in the confusion of daily life, forget why we wanted to do it in the first place.

To work effectively, we need to remind ourselves why we are doing what we do, and why it is important, and with that knowledge, continually rediscover our motivation. For some it will be working horses, for others, turning wood or making felt. The important part is nourishing the spark that inspires us, with thought, work and care.

Opposite - The author's horses lined up ready for the next furrow.

Chapter Five

Learning to Plough

Heavy Horse World, 2020

For many people, ploughing with horses is the very essence of horses working at their best. Compared with many other agricultural activities, ploughing requires a lot of energy, so when working with the limited power of horses, at eight to twelve inches each pass *(20-30cm),* it is not a quick operation. In the past, ploughing was therefore the main occupation of farm horsemen, so many people were very good at it, but nowadays it can seem like a skill so complicated and mysterious that it is difficult to know where to start.

There are several prerequisites to ploughing successfully; you need a workable plough, suitable land and horses which will walk where you want them to, at a slow to moderate pace. You also need to know the feel of the plough, how to adjust it and how to control the horses. Once you can do all these things, then you can plough, but for each one that you cannot yet tick off your list of competencies, the harder it will be to learn. Fortunately, some of these skills can be achieved before you ever get near a moving plough.

The Horses

When ploughing, the horses need to move at a steady walk, pulling a moderate to hard load. Many horses nowadays are not used to this, especially if they are mostly used in light wheeled vehicles, and just like when you need to push a heavy wheelbarrow, they tend to rush at a big load. If your horses have been allowed to do this, they are unlikely to pull steadily with a plough; they need to be trained to do it first. Regular field work with consistent moderate to hard loads is a good prelude to ploughing, but if you do not have that opportunity, training them in a sledge to which you can increase the load by adding weight, will also do the job. As you are doing so, keep them walking straight, pick a point in the hedge and aim for that. The horses also need to turn precisely and tightly on the headlands. If they have done land work, this will be normal part of life, but if not, make sure you turn them tightly when chain harrowing or when you are turning the sledge. Keep the traces tight whilst they turn, because when you are turning the plough, you need the contact via the traces to steer the plough. If your horses are unhappy with the feel of traces against their legs, then you need to go back to basics and in a safe environment, get them desensitised to this, and leave ploughing until they are unconcerned and have done a lot more work. Finally, the furrow horse needs to be happy walking down a furrow, the easiest way to learn this being to lead him down a furrow, and when he steps out of the furrow, put him back in with a suitable tone of voice, and praise him when he stays there. After that, drive him along the furrow on his own, then drive him together with his mate, keeping him in the furrow. The more steady work your horses have done, the easier it will be to get them to plough.

The Plough

To plough effectively you need a plough where all the parts work; the wheels should move freely without wobbling, all the bolts for adjusting the plough should move, and the hake at

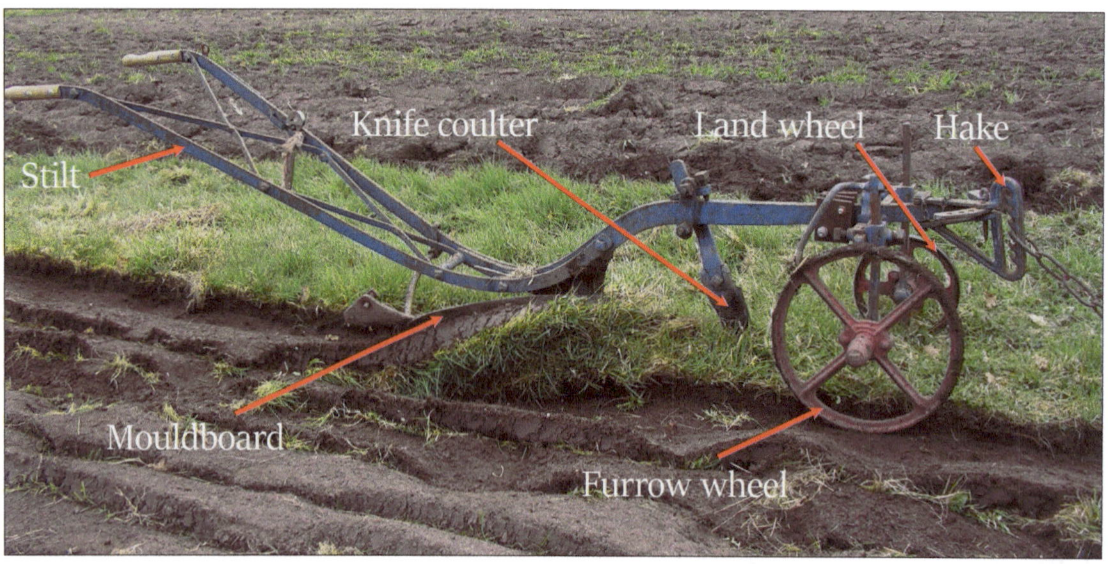

The parts of the plough, this time without a skimmer.

the front where the draught chain is attached should have its full range of movement and the means to fix it in any position. Unless it is a digger plough with a shin on the leading edge of the mouldboard, you need a coulter to cut the vertical cut, and usually a skim coulter to turn a mini furrow in front of the plough body. The point, or share, should be firmly attached to the frog, and all the wearing surfaces should be bright and shiny.

Unfortunately, horse-drawn ploughs have not been manufactured in Britain for seventy years or more, so finding a useable plough may be difficult. Finding replacement wearing parts, such as shares, may be even more difficult, so much so, that some models of plough are no longer viable as a useful tool. You are more likely to find parts for ploughs that were made by large manufacturers who turned out thousands of ploughs, especially if the same plough bodies were also used for tractor ploughs. Because many old ploughs are missing parts and have seized bolts, rusted and pitted mouldboards, cracked castings, and maybe a bent beam, buying an old plough is a risky venture, so until you know what you are looking for, ask someone who knows, or put off finding your own plough until you know by experience what to look for.

New ploughs are produced in America, and a few have been imported into the UK, though you can only get walking ploughs with short digging mouldboards. Unlike British ploughs, the American ploughs have only one wheel, so if you want to use them with two wheels, you will have to make some modifications.

The Land

Land varies in how easily it will plough. This is partly due to the qualities of the soil itself, whether it is sandy, silty, a stiff clay or a mixture, but it also depends on other factors. If it has been a dry summer, it will be hard for the plough to penetrate; if has been very wet, some soils can become very sticky, and for the sake of the soil and the horses, are better left until they are dryer. When you are starting, it is good to find somewhere which is reasonably flat, with an even surface, free of big stones and big weeds, such as a stubble field. If the land has been travelled by heavy machinery, especially when wet, then leave it alone. Over the last seventy years most land has been compacted by tractors smearing the soil, it is usually low in organic matter and low in worm numbers, all of which make it harder to plough.

Handling the Plough

Imagine you are standing on the headland of a partly ploughed field, the plough handles in front of you, ready to have a go, the plough in line with the next furrow to be cut. As the horses step forward, the big wheel, the furrow wheel, drops into the last furrow. As it does so, the point of the plough, the share, starts to cut into the soil and after a couple of feet the plough levels out, cutting the furrow to the correct depth and width. Since you are supposed to be controlling the plough, you grab hold of the handles and follow the plough. Although a well-set plough doesn't need much steering, at first it still seems that you are going at ninety miles an hour, because there is so much happening, even if the horses are drawing steadily. To begin with, you need to get used to walking in the furrow, and you can feel the plough's movement through the handles and see, hear and feel the soil being turned to the right. The turning furrow can be a bit mesmeric, so concentrate instead on steadying the plough handles, resisting any big movements.

As you get used to this, you will start to notice other things, the smaller land wheel up ahead to the left, travelling on the unploughed land, which helps to control the depth and keeps the plough stable, but don't worry about that for now. Pay attention instead to the furrow wheel in the corner of the furrow, next to the furrow wall. At times this wheel may move away from the furrow wall, so the plough cuts a narrower furrow, or sometimes it may press hard against the furrow wall as the plough tries to cut a wider furrow. To avoid this happening, you need to push or lean sideways on the handles. Steering a plough is like steering a boat with a tiller; to go to the left, you push to the right; to go to the right, you push to the left. The reason it works like this is because the plough pivots about its mid-point, so, for example, when you press to the left, you push the back of the landside against the furrow wall, pivoting the share away from the furrow wall, so it takes a narrower furrow.

Steadying the plough, resisting any sideways movements but not trying to push it. The hemp halter shanks from each horse's halter, tied to the other horse's hame ring, keep the horses walking at the same speed. In conjunction with the blue coupling band between the horses' bits, they also keep the horses the right distance apart. When Lucy, the darker horse, moves her head back to a straight ahead position there will be a similar amount of tension in the coupling band as there is in the plough cord. If the horses don't need to be held back, having some slack in the lines and working with gentle or minimal pressure on the bits is the way to go.

Unlike a boat, this topsy-turvy steering also works in the vertical direction, so as you press down on the handles, the plough pivots on the rear of the sole, the point is lifted, so the plough rises through the soil and the furrow becomes shallower. Lifting the handles, on the other hand, tends to make the plough sink deeper, but the amount is limited by the wheels.

As you get to the end of your first furrow, press down, particularly with your right hand, and the plough will come out of the ground. One furrow done. When you look back at your work, it won't look perfect. On some soils the plough runs true, but often there are changes in soil type over the length of the furrow which, as a beginner, you weren't ready for. So next time down the furrow, you might see that there is a wobble where the furrow wall goes to the left, where the plough took more land. When you get there this time, you need to lean to the left, pushing the furrow wheel away from the furrow wall so the plough takes less land, making the furrow straighter.

Adjusting the Plough

The Depth

In most ploughing, you are aiming to plough at a consistent depth, so the bottom of each furrow should be level. Since the furrow wheel runs in the bottom of the previous furrow, this wheel should therefore be set at the same level as the share. If you were to raise the furrow wheel from this position, the wing of the share would cut lower than the wheel, so the right-hand side of the furrow would now be deeper than the left-hand side. Similarly, the height of the land wheel can be raised or lowered to adjust the depth of the left-hand side of the furrow. (see illustration on the next page) In normal ploughing, if you want to change the depth of ploughing, you usually just adjust the depth of the land wheel.

The Width of Ploughing

The width of the furrow slice can be altered by moving the furrow wheel sideways on the cross piece. If you move the wheel to the left, nearer the share, the whole plough is moved to the right, so the plough cuts a narrower furrow. Move the wheel to the right, away from the furrow wall, and the plough can take a wider furrow. However, that is not the whole story. If the plough was set correctly and you move the wheel away from the furrow wall, say by an inch, when the horses go forward the wheel will still be travelling an inch away from the furrow wall, so the furrow will obviously remain at the same width. You could push hard to the right on the handles to make the plough take this extra inch of land, but it is hard work. Instead, you need to make another adjustment, this time to the hake.

The Hake

The hake at the front of the plough is usually a quadrant with a locating pin and a vertical piece of metal with a series of notches to adjust the height of the draught chain from the horses' eveners. The position of the hake determines where on the plough the force from the horses is focussed, and it is the prime method of regulating the furrow width. The adjustment of the hake works in the same back-to-front way as steering the plough handles, so if you want the

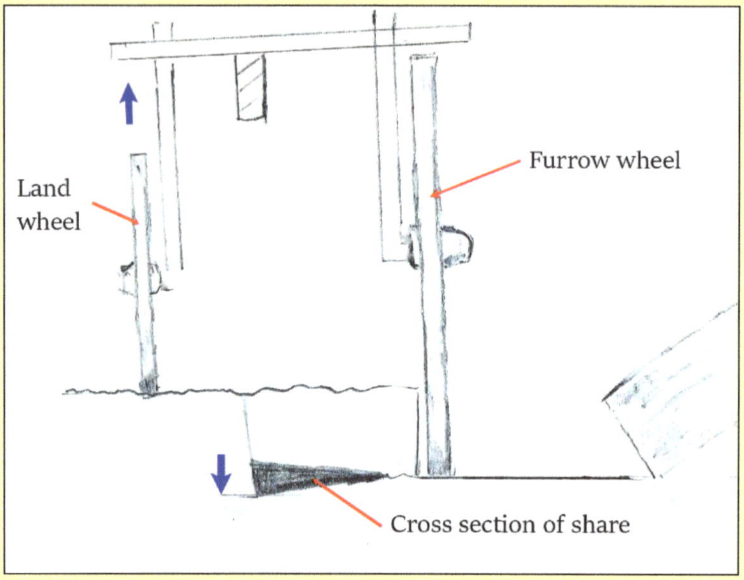

When the land wheel is raised on the cross piece, it allows the plough to tip over towards the land and the share to sink further into the ground so the left hand side of the furrow is cut deeper.

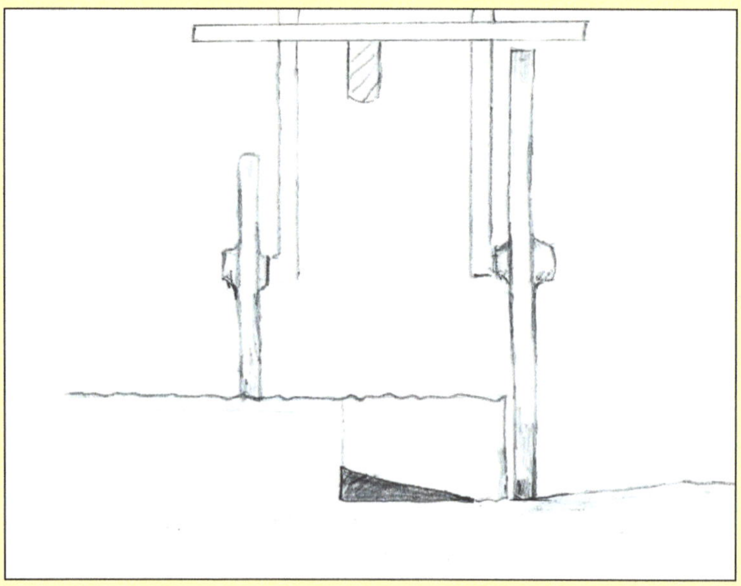

Ploughing the next furrow without adjusting either wheel. The bottom of the open furrow cut last time (to the right of the land wheel) is sloping, so the corner of the furrow where the furrow wheel runs is now deeper than before. The land wheel keeps the furrow at the new depth, and the plough now runs level.

plough to go more to the left and take a wider furrow, move the draught chain on the hake to the right. If you want the plough to go to the right and take a narrower furrow, move the chain to the left. If you want to plough more deeply, move the draught chain upwards. To plough more shallowly, move the draught chain down.

If you didn't have any wheels, adjusting the hake is all you need to do to adjust the plough, but wheels do make ploughing easier and prevent the plough from going too deep or too wide. However, you can have the adjustment of the hake at odds with the adjustment of the wheels. For instance, you may find that the furrow wheel is pressing hard against the furrow wall and trying to climb up it. This is because the draught chain on the hake is too far to the right, so the plough is trying to plough a wider furrow than the wheel is allowing it to cut. To remedy this, you can either move the furrow wheel out a bit, but this may make the furrow wider than you require, or you can move the hake a notch or two to the left which will allow the plough to run straight and the furrow wheel to turn happily near the furrow wall. Conversely, if you are continually having to push the handles to the right in order to get the furrow wheel close to the furrow wall, then moving the hake a notch or two to the right will correct this.

The vertical adjustment on the hake is slightly more complex. If the draught chain is hooked on too low, the share is tipped upwards, so the plough will plough more shallowly, and the wheels may come off the ground. Conversely, if the chain is fixed too high, it will tend to force the plough deeper, but the wheels will prevent it from doing so. Instead, extra pressure will be put on the wheels, the plough will tend to run on its nose and the draught will increase. To remedy this, once the plough seems to be set correctly, it is a good idea to lower the draught chain one notch at a time until the plough starts to lose its depth, then move it back up one notch. This will take most of the pressure off the wheels whilst still allowing them to regulate the depth, and the plough will run more easily.

When the hake is set correctly, the line of the traces and draught chain line up with the centre of resistance of the plough, minimising any excess draught.

The draught chain from the horses is at the lowest position on the hake which tends to pull the plough out of the ground. The lateral adjustment is central on the quadrant so the position of the plough will be equidistant between the horses.

The Coulter

The last adjustments on the plough itself are the coulter and the skimmer. The coulter makes the vertical cut of the furrow wall. There are two types of coulter, most traditional general-purpose ploughs having a knife coulter, but many can be used with a disc coulter instead. Knife or disc coulters should be set about ½" *(1.2 cm)* to the left of the share to leave a cleanly cut furrow wall. The tip of a knife coulter should usually be set a little in front of the plough point and a little above it. The disc coulter is useful if there is a mass of grass roots to cut through or if there is a lot of vegetation on top of the soil. To cut through grass, the disc should be set deep to cut through the majority of the roots, whereas with a lot of loose stalks or leaves on the surface, such as after mangolds or a green manure, the disc needs to be set quite high, so the crop residue is squeezed against the soil surface by the rotating disc coulter as it is being cut.

The skim coulter, or skimmer, does a completely different job from the knife and disc coulters. The skimmer is like a mini plough which turns a tiny furrow in front of the main plough. It removes the top left corner of the furrow slice, dropping the soil and vegetation at the bottom of the furrow. Its purpose is to eliminate the possibility of vegetation poking through the seams between the finished furrows, because it is this corner of the furrow slice that remains near the surface of the finished work. The skimmer should be set to cut the very edge of the furrow slice, and only as deep as is necessary to bury the trash.

With most of the furrow slice removed, the action of the disc coulter in cutting the vertical cut is clear, and the share has started to cut the bottom and lift the furrow slice.

Because the coulter is set to the left of the share, there is a gap between the front of the mouldboard and the furrow wall. The right-hand side of the furrow slice, in front of the worm, has not been cut by the share and forms a hinge around which the furrow slice is turned.

Horse Skills

Before ploughing on your own, you need to be competent, confident and relaxed with your horse driving skills. You should also know what will happen if one horse steps forwards, backwards or sideways, and what will happen to the implement, to the other horse, the lines, and the traces. If your horses end up at right angles to the plough, with two legs over traces, you need to know that your horses are calm, know what to do, when to sort it out and where the potential dangers are. If you are not that far advanced but want to have a go with a plough, get someone else with better driving skills, and ideally, knowledge of ploughing, to drive the horses while you learn the basics of controlling the plough.

Adjusting the Horses

If I took my horses out of a plough and you put yours in, the plough may not run the same. If your traces were shorter than mine or your horses were taller, then the angle of the draught chain will be higher so it will tend to pull the plough out of the ground, so to keep the plough running true you would need to raise the draught chain on the hake. Alternatively, the plough might cut a narrower furrow, because your coupling band is shorter, keeping the horses closer together, so with the furrow horse in the furrow, the line of draught is closer to the ploughed land. If you used a narrower cobbletree (two-horse evener) that would also bring the horses closer together and make the furrow slice narrower. In both cases you would need to move the hake to the right.

How to Learn

Although ploughing looks simple, it is a complicated business with many variables. Luckily, some of these skills can be learnt one at time. For anyone just starting out, learning horse skills must be the highest priority. You can also learn about ploughs and adjusting them by watching and talking to ploughmen, but the only way to learn how to guide the plough and to use the feel of the plough to tell you how to adjust it, is by ploughing. Unless you are lucky enough to work on a horse-powered farm, probably the best way to learn is on a ploughing course, or to apprentice yourself to a knowledgeable ploughman, though some may be more interested in doing the ploughing themselves rather than teaching a beginner. There will, however, be some people who want to give it a go on their own. If you do, start by leaving an open and shallow furrow from your tractor plough (or from your horse plough on a long chain behind a tractor) and teach your furrow horse to walk along it. Then attach the horses to the plough, put the furrow wheel in front of the share and the hake all the way to the left so the horses will just drag the plough down the open furrow without cutting new ground. When you have all got used to that, reset the furrow wheel and move the hake back to the middle and let the plough work. Having someone experienced to drive the horses will make all the difference while you concentrate on the plough. At the bare minimum, have someone who is confident around horses to stand at the horses' heads, or do anything else that is necessary while you adjust the plough. Nonetheless, you will be on a steep learning curve. No matter how you start, spending lots of time driving horses, and ploughing without having to set out a field will make the next part easier.

Dividing the Field

Most land today is ploughed with a reversible plough, which is in effect a right-handed plough with a left-handed plough mounted on top, the whole lot being swivelled over at each headland and used alternately on each run up and down the field. This allows the ploughing to start at one hedge and continue across the field, turning all the furrows in one direction. Most horse ploughs, however, have a single mouldboard which turns the furrow to the right, so if you were to plough along the hedge side, you would have to return empty to the starting point to lay another furrow against the first. To avoid this, the field is laid out into sections, into parallel strips called lands, which are usually 11 to 22 yards *(10 to 20m)* apart, with the headland all around the edge of the field to turn the horses. To set out the field, a very small furrow, a scratch furrow, is ploughed all the way round the field, parallel to the hedges, about 5 to 6 yards *(4.5 to 5m)* away. This is the line where you will put the plough into the ground and take it out at the end of each furrow. The lands are then marked out, each with a peg at both ends of the field. Then sighting from one end of the field to the other, another peg or two will be put in between, which is the line you plough to when ploughing your first furrows.

The Tricky Bits

Then comes the time to plough that first furrow. With the plough in line with the marker pegs for the first land, raise the land wheel so the point of the share will plough two to three inches deep, and the furrow wheel a little bit too so the wing of the share is just in the ground. The horizontal adjustment of the hake should be in the middle, along the line of draught, and the draught chain should be set relatively high to encourage the point to go in the ground. Now line up your horses so you can see between them along the line of the pegs and drive as straight as you can, keeping the pegs in line. This is where your practise driving other implements in a straight line will pay off. Don't turn around until you get to the other end, and when you do get there, try not to worry about what you see! Turn the horses to the left to face back down the field, putting the horses either side of the new furrow. You now want to plough another shallow furrow away from the one you have just made, so they leave a shallow V-shaped trench. With your horses either side of the existing furrow, you will see that the bit of land you want to plough is not equidistant between the horses, but more behind the offside horse. So, set the hake a few notches to the left. You also need to think about where the little wheel is running. It needs to run in the open furrow, not on the bumpy turned furrow, so should be moved nearer the plough's beam. The depth may also need adjusting because you want the left-hand side of the share to cut a little deeper than last time, so that the landside has a ledge of soil to run against. This will prevent the pressure of the turning furrow from pushing the plough sideways. Once you start the horses, if you need to push hard one way or the other to keep the plough where you want it, stop and adjust the hake.

At the end of the furrow, turn to the right and put your offside horse, the furrow horse, into the furrow. You are now going to turn those two shallow furrows back towards each other, except deeper than before to form the crown or rigg. Move the land wheel to the left to give the plough more stability and reset the height of both wheels to allow the plough to take a deeper furrow. Also drop the draught chain a couple of notches on the hake. At this stage you don't want to plough to the final depth, otherwise the crown will finish up higher than the rest of the work. Unless you are using a plough with a wide body or have a narrow cobbletree, the hake will probably need to be a few notches left of centre. After a couple of yards, you may find you need to adjust this horizontal adjustment on the hake for the plough to follow the correct line. Once you have ploughed these two furrows back against each other, you have ploughed your 'top,' 'crown,' 'rigg' or 'ridge' and hopefully it will be reasonably straight.

Now you can set the land wheel deeper to the final depth for normal ploughing. Unless you want the wing of the share to cut a little deeper this next pass, you also need to set the furrow wheel at the same depth as the share. This can be done by sighting down the landside and wing with the plough on its side, though once the plough is in the ground, you may have to refine this adjustment.

An alternative way to plough a rigg, more suited to stubble than grassland as the soil in the middle of the rigg is not moved. In this case, all the furrows are ploughed towards each other.

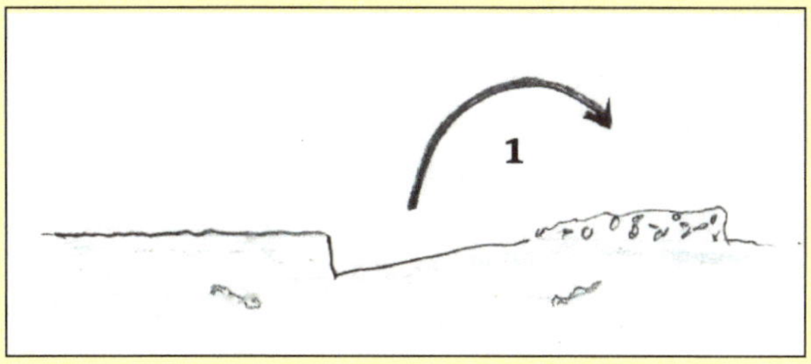

The first shallow furrow, as seen from behind the plough.

The view from the same place, the second shallow furrow having been turned away from the first, but just a little deeper to leave a step for the plough slade to run against to keep the plough running straight.

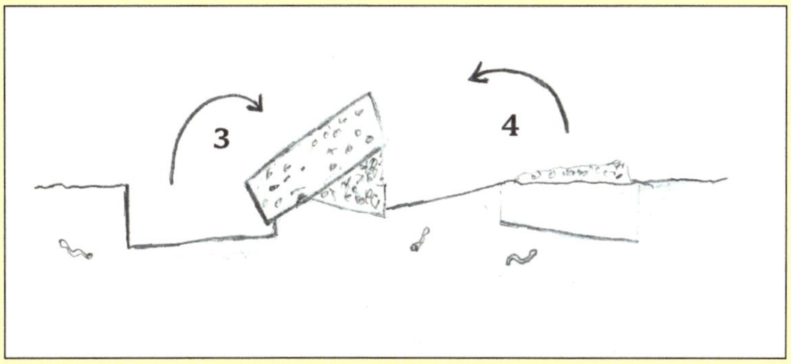

The third furrow turns the second furrow back on itself whilst cutting deeper into the land. The fourth furrow does the same to the first furrow to complete the rigg.

Turning the horses on the headland requires balancing the plough whilst steadily turning the horses just the right amount to line up for the next furrow.

Continue ploughing round the rigg, turning to the right at each end. Correct for any wobbles by steering the plough to take a narrower furrow where there is a bulge along the furrow wall. With a big wobble you might have to do this over a few turns; don't try and correct it all at once. Stop ploughing when the ploughed strip measures half the distance between the riggs. So, if the rigg marker sticks are twenty yards apart, you plough a strip ten yards wide, five yards each side of your marker sticks.

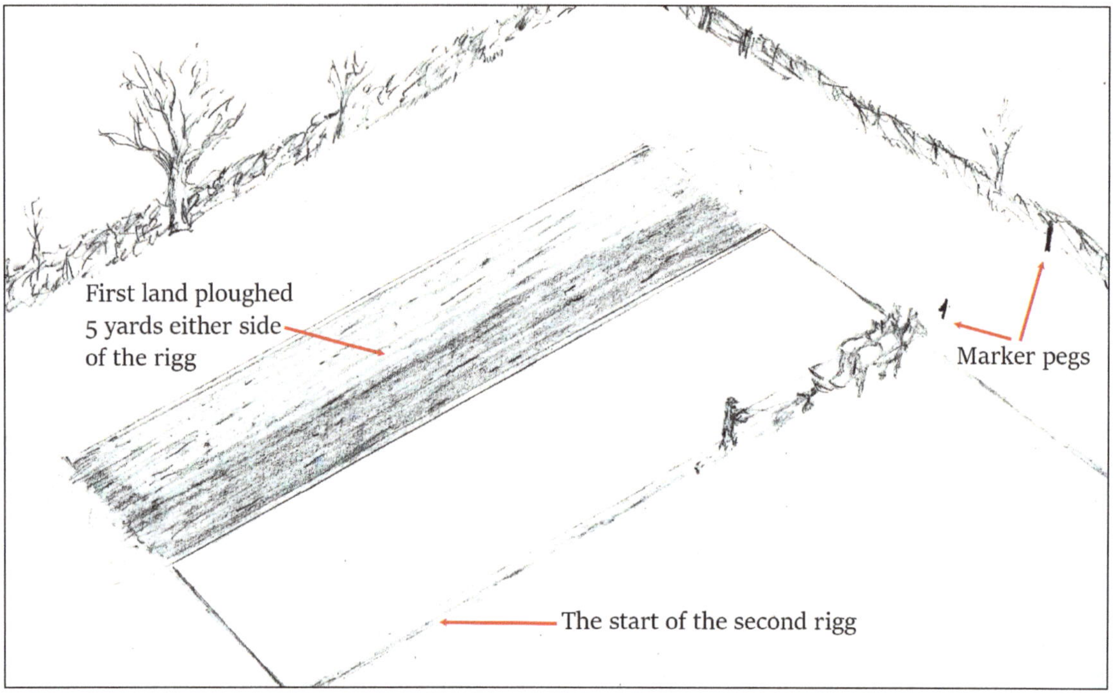

Then you go to your next land and plough another rigg until it is also ten yards wide. This leaves a ten-yard strip of unploughed ground in between. Plough this part turning left at the ends. As the unploughed ground becomes narrower, measure it at both ends by pacing it heel to toe, to see if it is parallel. Often it won't be. To avoid leaving a long thin triangle at the end, correct for any inaccuracy by narrowing the furrows at the narrow end.

The Finish

If you were to continue to plough at the same depth as you come to the finish, it would leave a very deep furrow which would be inconvenient to cultivate afterwards. So, for the last yard or so, lower the land wheel a little on each turn, so the plough ploughs less deeply. At this stage you might also need to adjust the width of the furrows, to ensure that the last furrow is the full width. Besides looking tidy, the full-width furrow turns over better than a narrow one. As the land gets narrower, the land horse will have to walk in the other open furrow. Because this may put the horses further apart, you may need to change the length of the coupling band between the horses and/or adjust the hake to allow the plough to run where it should. On the

As the unploughed land gets narrower, the land horse needs to walk in the other furrow, which may require the coupling band to be lengthened and the hake to be adjusted to the left so the plough does not take too much land. On the next furrow, the hake will need adjusting again as the horses, still both in a furrow, will be closer together. Subsequently, the land horse will need to walk on the ploughed land.

last furrow or two, the land wheel will have no unploughed land to run on, so to maintain the right depth you need to set it deeper to run along the bottom of the other open furrow. When you have only one furrow to turn, the land wheel needs to be raised a little to allow this last furrow to be ploughed a little deeper. This is to create a ledge of soil, so the landside can resist the sideways pressure of the turning furrow. The very last furrow, the mould furrow, is taken out of the bottom of the open furrow, and if you can leave it dead straight, everyone should be highly impressed. Having made the last few furrows shallower, it seems surprising that you might want to take an extra furrow, albeit a small one, at this stage. Its name, the mould furrow, suggests its purpose, that is, to create some loose soil, or 'mould,' which gives something for the harrows to go at and the seeds to grow in.

The Headlands

Once the centre of the field is ploughed, all that is left are the headlands. These are ploughed round and round the field, either turning the furrows inwards towards the existing ploughing or outwards towards the hedge, usually alternating each year so the field is kept level. When starting from the inside, the first furrow needs to be shallower than usual, and will be a bit challenging, especially if you weren't precise in getting the plough in and out of the ground exactly at your scratch mark. This is when you need to keep pressure on the plough handles to keep the plough running steady, to avoid the wheels dropping into a furrow that went past the scratch mark. Once that first headland furrow has been made, the plough can go deeper. After a few turns around the field, the horses' heads will reach the hedge before the plough has reached the point to make a 45-degree corner in the ploughing. To turn the ground the plough hasn't reached, as you turn the horses through 90 degrees at the corner, pull the plough back before starting along the next side of the field so the missed ground gets ploughed. Nonetheless, after a few more circuits of the field, there will not be enough room for you and the plough, so the very corner will not get ploughed. Before the First World War when labour was plentiful, precise farmers would send someone with a spade to dig the corners, but nowadays it is perhaps best to think of it as a tiny wildlife corner.

When ploughing the headlands the other way round, turning the furrows towards the hedge whilst going anticlockwise around the field, a shallow furrow is first turned towards the hedge, the next furrow being almost to full depth. The corners present the same problem as when ploughing towards the middle, but finishing is easier, because the furrow wheel will have a nice level furrow to run in, even as you get to your less-than-precise ins and outs.

Delight

Ploughing with horses, when it is going right, is a delight. When it isn't going right, if you have horses that go too fast or won't stop immediately, or if you can't work out how to adjust the plough, then it is not so fun. Ploughing requires a range of skills, good horse skills in particular, just to make it work at all, but once a basic competency is achieved, the refinements of ploughing are a challenge that many people enjoy.

Opposite page - The author harrowing with four of Geoff Morton's good horses.

Chapter Six

Good Horses

Heavy Horse World, 2018

It was over thirty years ago that I drove a horse for the very first time. I still remember the strange but excited feeling of having this great animal ahead of me, knowing that I had to try to get him to move, the only connection between him and me being the hemp cords from his bit, passing through rings in the back band to my hands. Being a well-trained horse, I got him to start easily enough, then, intending to steer him to the right, I pulled on the right cord; he turned to the right, much too sharp to the right, oh dear, and walked back down the tight cord straight towards me! I took hold of his bridle, and he came to a gentle halt. Luckily, I've improved a bit since then, but from that day on I learned to appreciate good horses.

My early training continued with driving a dozen other horses, horses that knew what to do in most situations, because unlike me, they had experienced most situations. Although it would be wrong to downplay the value of the instruction I was given, in many ways it was those good horses that taught me what it is to be a horseman.

Spending days working with horses, I have been in the right place to witness many natural events, barn owls patrolling the hedgerows, kestrels plummeting to catch prey exposed by the mower, and frogs hopping out of the way of the horse-drawn tedder. I have ploughed, cap

pulled right down, in freezing rain, longed for the half-hearted sun to burn away the fog of a chill October morning, and, standing on a load of wheat sheaves at the end of a long day's harvesting, welcomed the long shadows creep over the harvest field, casting elongated images of the horses and stooks against the stubble. In all these pictures, the horses were the central focus, sometimes standing like statues, more often moving forward with purpose, clean work-hard shoulders leaning into well-fitting collars, ears forward, at ease with their life.

To a bystander, it all makes a picturesque scene, a camera-worthy image to try and capture the magic. Being part of that picture, however, is altogether different; it is an all-inclusive experience, as you focus on to the horses' movements, watch their ears, glance at the traces and swingletrees, listen to the implement, feel the soil beneath your feet and their mouths through the lines.

The magic of doing work with horses is in the doing; you can't capture it with a camera because it is fleeting and dynamic. Some of the magic stems from working with natural processes, working outside the normal human realm, from doing something which is hard yet is also easy, restricted by the limitations of the horse but broadened because you are using horses. But mostly the magic lies in the relationship between the horses and the human, sharpened by the need to do a job and the cooperation necessary to accomplish it.

When we talk about the relationship between man and horse, most of the time we focus on the horseman being in control. Indeed, I would be amongst the first to criticise when someone doesn't give a horse clear instructions or does not insist that their horse stops on "whoa." But our

The author's first time operating the binder with Swale and Tees, two exceptional horses, under the watchful eye of Mark Morton.

relationship with horses at work is more complicated and subtle, because the communication goes both ways. It might only be the fact that you choose to stop when you can see that the horse wants to stop anyway, though conversely, you may urge him on because you don't want him to stop at that moment. Or you may permit a pair of plough horses not to start the instant you say "giddup," because you know that after they have looked at each other they are going to start simultaneously and smoothly in three seconds' time. Nor will you mind if they suddenly stop, because there may be a good reason.

On one occasion when this happened to me, I was tedding hay, driving clockwise round the outside of the field, when Molly came to an abrupt halt. She had just noticed the manure heap (which she had put there, but perhaps she had forgotten!) or perhaps it was something near it. There seemed to be nothing to worry about, so I clicked my tongue and off we went again. On the next round, this time driving anticlockwise, she stopped at the same place for no apparent reason. "Go on," I said a bit crankily, she took one more step, then stopped again. This provoked me to have a proper look, because although I was keen to get on with the hay, I know that when there is anything wrong, the place to fix it is when everything is stationary and calm. So, unless it is an avoidance mechanism, I don't mind a horse that stops. Anyway, my better look, which must have taken all of two seconds, revealed that the back chain had come undone, so the shafts were too low and were hanging on the shoulder chains. Molly knew this didn't feel right, and I learnt that I should have paid attention to her one step sooner.

Although this example is scarcely earth shattering, on other occasions I have had traces come undone, buck rake tines stick in the ground and threaten to break, and other instances where something has not felt right. Often the horse's decision to stop has prevented something worse from happening. Of course, sometimes when they stop, they are wimping out when a load gets hard. This is where the horseman's judgement comes in, deciding whether there is in fact something wrong, or whether the horse just needs a bit of encouragement, perhaps with a change of direction to make the load easier to start.

Once we acknowledge that horses have greater capabilities than merely following orders, we allow ourselves to reap the benefit of their contribution. I became acutely aware of this possibility within a couple of months of starting to work with horses. I was harrowing a field with four horses, a young Ardennes/Shire cross gelding on the near side, a big Clydesdale gelding called Swale next to him, and two reliable Shires on the offside. We harrowed all morning, then after lunch I drove the horses back to the field, driving diagonally in front of the harrows to get them into position to yoke them up. I stopped them, then attempted to swing them round to line up with their own swingletrees. The three older horses were happy to do this, the offside pair stepping back a step, but they couldn't finish the manoeuvre because the little gelding wouldn't move his back end over, he had gone into stand-by mode, happily standing still, paying no attention. Now I can think of three remedies to the situation, but back then my efforts proved unsuccessful and I started to get quietly frustrated. Then very deliberately, Swale slowly lifted his back leg and gently kicked the little gelding, who did a deft sidestep which put him directly in front of his own swingletree. I could then move the others into position, I yoked them up, and off we went.

You might think that it was just a coincidence that Swale did what he did, but I think he knew exactly what he was doing; he knew what needed to happen, and he made it happen. Having written this down I must admit to feeling slightly awkward; it is not because I worry about admitting to a lack of skill or fear being seen as a romantic; it is because those with extensive horse experience will have similar stories, perhaps more impressive stories, as this is just the sort of thing that happens when you spend enough time with horses at work. It is at once normal but also extraordinary. Most of the time, however, the joy of working horses comes from smaller things, the low nicker as you come out of the house, seeing a young horse settle into its work, or on a cold morning, watching steam rise from the horse's back as a shaft of sunlight bursts through the trees. Or like me, you may get a gentle buzz from seeing your equine companion dip her head through the collar, or, with just a word, push the cart back just one step.

In themselves these are all seemingly insignificant events, but they don't magically appear out of nowhere. They are developed over time, through good working routines, gentle persistence, observation and collaboration. Although each experience may appear unimportant, each tiny lesson reinforces or refines what has gone before. It is the layering of such experiences, building confidence and acceptance, which results in good horses, and, providing you have been paying attention, good horsemen too. The relationship between a horseman and the working horse is undoubtedly a remarkable thing. For me, thirty-something years on, it is still as intriguing and fascinating as it ever was.

Above - Gerhard Aschoff's Rheinisch-German gelding, with ears forward, contentedly keeping to the rows at the Pferdestark event in Germany.

Below - Stephen Anderson-Jones uses heavy cobs on his farm on the west coast of Scotland, seen here tedding hay using a home-made hitch cart with an engine.

The late Roger Clark expertly and seemingly effortlessly guides his horses whilst splitting ridges to cover seed potatoes. The Percheron is walking on the ridge, a challenging change for a horse used to walking in the furrow. On the next pass it will be the Suffolk's turn to walk on the ridge whilst the Percheron will follow the newly-made furrow.

Above - Shortly before this picture was taken, these horses were a litlle agitated. Now, under Robert Sampson's calm and purposeful guidance, they are settling into their work, the nearside two still alert to everything around them, the offside horse observant but just getting on with it.

Below - Daniel Hoffman's horses waiting to go into the demonstation ring at Pferdestark near Detmold in 2015. Their relaxed attitude is a reflection of the horseman's demeanour.

Heike Veidt and her two decorated Black Forest Coldbloods at the start, and nearing the end of her ploughing at the 2013 Pferdestark. The design and decoration of harness, as with the breeds of work-horse, vary widely across Europe. What good horses have in common is their ability, both physical and mental, to do the work.

Above - Whilst in motion, Marty Gilles adjusts his reversible plough, pulled by an Ardennes and a Breton draught horse. Below - Horse logger Frankie Woodgate and her Ardennes geldings moving logs with a forwarder. Especially when working as a single, horses are very manoeuvrable and their minimal impact on the soil is beneficial both for the soil-dwelling organisms and for the trees.

Chapter Eight
Ploughing with a Single Horse

Small Farmer's Journal, 2013

If you look at old photographs of horses ploughing, most of them feature a pair of horses. If photos had been invented before the 1860s there would have been a greater variety of hitches, sometimes with three abreast or three in line walking down the furrow, three yoked bodkin fashion (with two in the furrow and one on the land), or fours hitched in line or in two pairs. In the 19th century there were significant improvements in plough design, so that farms where four horse teams had once been used, could now plough with three or just a pair. By the time photographs became more common in the early twentieth century, two horses could pull the plough on all but the heaviest land, and this became the most common hitch. By comparison, images of a single horse pulling a plough are relatively rare; perhaps those who could only afford one horse could not afford to have their photograph taken either, but all the big plough manufacturers in Britain produced light ploughs for a single horse or a pair of ponies, and there were many smallholdings which only supported a single horse.

Even though ploughing with a single horse is possible, using a pair has many advantages. Once the furrow horse knows that he must stay in the furrow and the land horse is the right distance away from him, guiding the horses almost becomes unnecessary. With the horses the right distance apart, the line of draught through the cobbletree is pretty much in line with the centre of resistance of the plough, so the plough requires the minimum amount of force to pull it. The plough is also stable in the ground, making it is easy to steer and adjust. With a pair of horses, you can also see between them to keep the furrow straight or to even out any irregularities, and their combined power is well matched to the size of furrow.

Compared with using a pair, ploughing with a single horse presents additional challenges. Depending on the plough used, the nature of the land and the crop, the pull is substantial. Indeed, some people say that ploughing with a single horse is too difficult, though the existence of the old single-horse ploughs contradicts that opinion. Nevertheless, there are some conditions, such as heavy clay soil or established grass which could be beyond a single horse's capabilities. With a single horse, there is also the issue of where the horse walks, either on the unploughed land in front of the plough, or in the furrow. Theoretically, having the horse on the land means his efforts are put to better use, but with the horse in this position, you can't see where you are going.

When I started out ploughing with a single horse on my own place, I already had experience of ploughing with two and three horses, but Molly had never ploughed before. However, she had done plenty of other work including pulling significant loads at a slow speed, and, most importantly, she will stop and stand while I attend to an implement. The single horse plough I have is a Ransomes, Simms and Jefferies RHA, a light general-purpose plough, though I had neither skim coulter nor disc coulter when I got it. Although it is quite different from the Ransomes LCP ploughs that I had used previously, which are much more heavily built semi-diggers, the adjustments are all the same.

The RHA plough, with a general purpose mouldboard which turns the furrow intact, seen here ploughing at only two to three inches depth.

To give myself an easier start, I first got my wife to drive a tractor dragging the horse plough to make the first furrows while I steered the plough. It was then I made my first error by putting the horse on the land in front of the plough. Although this seemed like a good idea, I soon discovered why all the old photographs show the horse in the furrow. The problem was that when my mare's feet came down just next to the furrow, the furrow wall collapsed under her foot, which not only left a lump for the furrow wheel to climb over, but because she didn't like the ground giving way, she tended to wander away from the furrow edge. This made it impossible to keep the plough going straight, and sometimes it came right out of the ground. Once I had changed my mind about where she should walk, I drove her a few rounds in the furrow without pulling anything, then I got her to drag the plough down the furrow without cutting any soil so that she was happy staying in the furrow before we went back to ploughing.

Once she was in the furrow, the plough did not wander and I could see where we were going. When the horse is in the furrow, however, the pull from the horse is significantly out of line with the centre of resistance of the plough, resulting in increased friction and some of the horse's efforts being wasted. To reduce the the angle between the horse and the plough, I added an extra foot *(30 cm)* to the draught chain, which is a trick used by match ploughmen when they do high cut work (oat seed furrow). They do it so that both their horses can walk far off on the land, so the furrows are not despoiled by feet marks. In their case they use a much longer chain, again to minimise the angle of draught.

The RHA ploughing grassland. With the horse in the previous furrow, you can see where you are going and the horse knows where to walk, but its pull is significantly out of line with the plough.

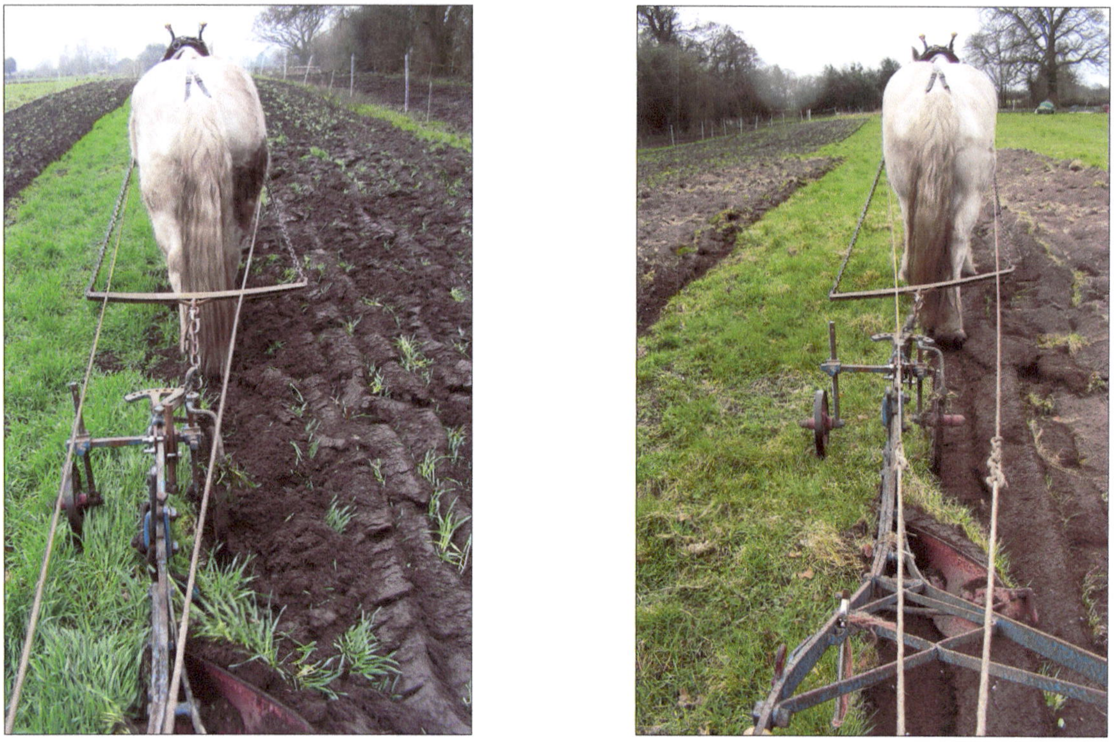

Left - Ploughing in a cover crop of rye, the furrow being 7" wide and 5" deep (18 x 12.5 cm). The draught chain is well over to the right because the rye provides very little 'bite' to the plough. Right - The draught chain is more central on the hake when ploughing grass, as even when only ploughing shallowly, the force of the firmer soil and grass roots on the mouldboard holds the plough more in line.

On land already under cultivation, my big horse and little plough worked pretty well. In comparison with a semi-digger plough pulled by two horses, it ploughs a small furrow, but on the small scale that I work, that is just fine. The bigger challenge was to tackle established grassland, a job which might prove too much for my horse. To give her a better chance, I thought I would first rip up the top couple of inches before ploughing. If I had had a set of discs suitable for a single horse, I would have used them, but in their absence, I tried repeated passes with my spring tine cultivator. It probably pulled harder than a plough, but even with many passes it made little impression on the grass. Then after re-reading one of Anne and Eric Nordell's articles,[1] I decided to experiment with skim ploughing a shallow furrow of only two or three inches *(5 to 8 cm)*. At first, I tried using a disc coulter which takes less power than a knife coulter, but at only two or three inches depth, the plough had too little weight of soil above the mouldboard, which, in conjunction with the upwards pressure created by the disc coulter, sometimes pushed the plough out of the ground. So, I reverted to the knife coulter, which, due to the angle it makes with the soil, helps keep the plough down. Even with the knife coulter, the plough does not hug the ground as well as when ploughing more deeply, so if you don't pay attention you can end up with it on the surface. To prevent this from happening requires a good feel for the plough, but the biggest assistance is having a plough with wheels.

When skim ploughing the top couple of inches of soil, all the adjustments on the plough are more crucial than when ploughing at a normal depth, but despite cutting through a lot of roots, skim ploughing turned out to be relatively easy work for the horse. Working at this depth, shallower than the plough was designed to work and without a skim coulter, the furrows are turned completely upside down and the mouldboard doesn't push the furrow slice down. Nonetheless, having the sod completely inverted means that regrowth is minimal, especially if the grass has been grazed down before ploughing. Although inverting sod to a normal plough depth is not necessarily the best thing for getting it to decompose quickly, at two inches *(5 cm)* down there is plenty of biological activity. However, ploughing at two inches is not really deep enough for a seed bed, so another couple of stages are necessary.

My next step is to roll the ground soon after ploughing, which pushes the furrows down tightly next to each other, and then broadcast mustard seed on top. Because it is very quick growing, the mustard can smother any grass which comes up between the furrow seams whilst using up the nutrients from the decaying grass. To cover the seed, I use a spring-tine cultivator set very shallow, though a disc would be better. The reason for using the spring-tine cultivator rather than a harrow is that it does not grab hold of any loose furrow slices, the tines springing out of the way if they meet too much resistance.

Covering the mustard seed with the spring-tine cultivator set very shallow.

The tines also explode bits of soil from the upturned sod, throwing them in the air which covers the seed, even when the ground is a bit dry. Admittedly, this is a rough way of making an imperfect seed bed, and if it is very dry the germination can be poor, but in average conditions the mustard soon produces a thick, weed-suppressing stand of foliage, severely curtailing any regrowth of grass. When the mustard is in flower, I mow it with my McCormick Deering mower, first with the cutter bar in its partially raised position, and then again with it on the ground. Then I plough the ground again to about five inches *(13cm)* in depth, this time with the disc coulter to cut through the mustard stalks and a skim coulter to bury the trash. This deeper

The mustard in flower, having suppressed nearly all the weed growth.

ploughing is also well within the horse's capabilities, once the grass roots which hold the soil together have mostly rotted away.

Doing the ploughing in two stages is twice the work of ploughing once, but I like ploughing and only have small patches to work, so in some ways having to do the job twice is a benefit. If I was using my horse more, was ploughing more, both on cultivated ground and grassland, and therefore had a very fit rather than moderately fit horse, I could, and sometimes do plough grassland where the soil is relatively light all in one operation.

The most difficult part of ploughing with a single horse is to get the first furrow straight. Whereas you can see between a pair of horses, the easiest way to drive a single horse in a straight line is to look right over the top between the horse's ears, which is obviously not an option when ploughing. Sighting between the horse's legs is perhaps another option and is what I do when following a scratch mark when making potato ridges with a ridging plough, but often you can only see as far as the front feet. When ploughing the first furrow, I now walk beside the plough with just one hand on one of the handles and sight down the side of the horse, driving with one hand, as straight as I can. Although I can feel whether the plough is wobbling about, I can't do much to change it. Instead, I rely on having the draught chain along the line of draught to keep the plough running straight and set the wheels wide apart to keep the plough stable. Although this works reasonably well, the results are not as good as when setting out with a pair of horses.

For anyone used to ploughing with two horses, all this effort might seem a bit excessive, but in practise, as long as the horse is in the furrow, ploughing with a single horse is relatively straightforward. Although the dominant culture today would have us believe that working with horses at all is hopelessly slow, inefficient and outdated, and that any physical effort is to be avoided unless it is recreational, those who work horses know this not to be the case. Given good horse skills and appropriate equipment, you can achieve much more than is generally thought possible. The same is true with the single horse, the weakest link still being the equipment, but I can imagine a time when a plough will be available with a rolling landside, wheels with roller bearings and with the option of a semi-digger mouldboard, which would require considerably less draught than my plough. The question as to whether you can or should plough with a single horse would then be answered once and for all. I may be jumping the gun, but I have already reached my own conclusion!

Chapter Eight

You Can't Work Horses and Send Men to the Moon!

Small Farmer's Journal, 2016

When I first got involved with working horses on Geoff Morton's farm, with the enthusiasm of the novice I was constantly asking questions. What I often thought was a straightforward question would often cause Geoff to pause, and as often as not the answer was preceded by -

"Well, there's more about it than you might think."

On one occasion I remember asking how many horses it would need to pull a particular set of discs. The answer was fairly typical for Geoff and went something like this –

> "Well, it all depends on what you want to do, and what land you're working on. On this light land probably two horses would manage it, but if you had the horses and were working long days, you'd want three, or if they weren't very fit or you were on strong land, maybe four. Then you have to consider the angle you set the discs at and how much they're sinking in, because that

> can make an awful lot of difference to how hard they pull. Really, these horse discs were made for working land after it had been ploughed because they aren't all that heavy, but nowadays when we use discs, we use them instead of ploughing, along with cultivators to break up the land after harvest, and that's a completely different job. When we use six or eight horses on tractor discs, of course, the first time over they're walking on solid land so it's easier for them, and you probably have the discs set straight to cut into the ground which makes it a bit easier too. Then second time over, with the discs set an angle to move the soil, it's a lot harder work for them so you might want eight horses on, whereas you'd have managed with six the first time over. Of course, when the ground conditions are right, they do a good job, and you cover a lot more ground than when you're ploughing...... It's the same as with all these things, if you have got good horses and know how to use them, you can do a lot of work with them, and do a good job too, but it's not the same thing as all this big tractor machinery and all the power that they have. You can't farm with horses and send men to the moon!"

For someone particularly interested in the day-to-day practicalities of working with horses, and with a fairly literal turn of mind, this last twist in the conversation took me aback; after all, I was a child when the first man went to the moon, so the idea was normal for me, certainly more familiar than doing farm work using horses. But there I was in my late twenties, working on a horse-powered farm, so that was obviously possible too. Of course, what Geoff was really saying was, although you can farm well with work horses, it is not for someone who is always in a hurry or who wants to get rich quick, or, in a wider context, it is not a way to create enough surplus to afford the heights of extravagance, such as sending people to the moon. He might equally have said you can't be a beekeeper and run a private jet, or a carpenter and live in a penthouse apartment, but by choosing the example of working with horses and moon travel, he hit upon what many people consider the epitome of backwardness, contrasting it with mankind's most impressive technological achievement, the triumph of man over nature.

Although that might be a common assumption, it does not take many moments to realise that the massive quantity of materials used to get astronauts free of the Earth's gravity are all products of the Earth, and no matter how big the fanfare for shiny technology may be, whilst the astronauts are in space their hold on life is tenuous. Then when they have eaten all their food, which was originally grown in some farmer's field, they have to come back down to Earth.

Nonetheless, the concept that we humans can, and should, escape the hold which the Earth has upon us, has not diminished since the first men landed on the moon. The new presenter of a scientific programme on the BBC, for instance, recently promoted the idea of people living on Mars as a solution for our overcrowded planet, and NASA is working on the technology to enable astronauts to do just that. However, there is never any mention of the high radiation levels on Mars and the storms of carcinogenic dust which block out the sun for more than a year at a time. Then there are far-distant planets that may be a bit like Earth, which are also suggested as possibilities for human habitation as we exhaust our own planet's resources. If some future technology did allow us to travel deep into space in search of a new planet or other life, one can imagine some intelligent alien life form in some far-off galaxy, after hearing of Man's plight, asking the space travellers whether they had ever thought of looking for signs of intelligent life on their own planet first!

To learn more about the processes necessary to sustain human life, in part to enable long distance space travel, an experimental project was started in Arizona in the late 1980s. Named Biosphere 2, this experiment involved a series of interlinked giant greenhouse-type structures, sealed from the outside world, each containing areas mimicking natural ecosystems, along with areas for people to live and to grow food. For two years it was occupied by six people, the whole project attracting huge attention and much criticism. The biggest criticism was that in this supposedly sealed and self-sustaining environment, additional oxygen had to be pumped into the biosphere at one stage because there had been a serious and unexplained drop in oxygen levels, preventing the people from sleeping properly or functioning effectively. It turned out that some of the carbon dioxide breathed out by the people and other organisms was reacting with the relatively new concrete walls to produce calcium carbonate, (limestone) so the oxygen contained in the carbon dioxide (which would usually be released into the air by the plants) was no longer available. Another problem was the extinction of many species, and the explosion in numbers of others such as cockroaches and ants which they initially did not even know were inside the bubble.

Perhaps the greatest achievement of Biosphere 2 was the growing of food. With limited space, the food growing area was like a domestic garden, relying on mixed cropping and a considerable amount of work to grow and process the food. In such a tight closed system where anything applied to the plants might well be eaten by the people within a week, the use of synthetic chemicals was avoided. Instead, they used biological and mechanical controls alongside the natural processes within the soil to provide nutrients. Nonetheless, the press was scathing of the venture; after all, because of the oxygen issue, it did not succeed in sustaining life as had been planned, but I suspect the root cause of the antagonism stemmed from the fact that the experiment fundamentally confounded the prevailing assumption that we humans are separate from and superior to nature, and as such, we can do anything we like. For those who believe that science can solve all our problems, it is profoundly uncomfortable to confront the reality that we are not as clever as we think we are.

Fortunately for us, nature in the outside world is much bigger and is capable of absorbing and straightening out many of our mistakes, although we seem intent on pushing that capacity for forgiveness to an alarmingly arrogant degree. There are, however, other experiments, small scale and down-to-earth, which show how working with natural processes brings benefits well beyond the immediate sphere of activity.

One such experiment is the work of a group of neighbouring farmers from the Pontbren river catchment area of mid-Wales, [1] who came together to work out how they could do something about their falling profits whilst securing their farms for the future. On their upland, marginal farms with heavy rainfall, sheep were the main enterprise, and following the usual trend to push for greater output, they were using cross-bred lowland sheep rather than the hardy traditional breeds. Although the farmers were getting more money for each lamb sold, the local environment is so severe that they needed to buy in more feed and keep the sheep inside for longer, whilst the prices for everything they bought were increasing faster than their incomes. To decrease their costs, they decided to go back to the tougher hill breeds so the sheep would need less feed and could stay outside more, but to provide enough shelter they knew that they would have to improve and renew hedges which had been neglected or pulled out in the drive for greater 'efficiency,' the hedges running along the contour having been the first to go. Over the following years, the farmers reinstated hedges and planted new shelter belts, with woodland strips alongside all the streams, often working together and buying fencing materials and saplings collectively. Unlike when planting for timber production, the trees were planted close together, with a high proportion of shrubs to slow down the wind, which in wintertime often carries the rain horizontally. The new plantings increased the wooded area from 1.5%, most of which was neglected, to 5%, only decreasing the area of productive land by a small amount. The effects of the Pontbren farmers' actions have been many; the provision of shelter has indeed proved positive for the livestock, there has been an increase in wildlife, and the farmers have discovered a sense of community through overcoming difficulties and working together. Other outcomes, however, were not expected. Perhaps most significant has been the change in movement of water through and on top of the soil. This was noticed by the chairman of one of the partner organisations, Coed Cymru (Welsh Woodlands), who, during particularly heavy rainfall, noticed how the water ran downhill in sheets across the surface of the tightly sheep-grazed fields, but this movement stopped at the hedges. The Pontbren experiment prompted a series of university studies, which, amongst other things, showed that the wooded areas have sixty times the capacity for absorbing water compared with the adjoining grassland, and that this effect starts to occur after only two years. They also found out that in times of particularly heavy rainfall, the peak flow of water in the river was reduced by a quarter.

A similar phenomenon was observed in Yellowstone National Park, where since wolves were reintroduced, there has also been a reduction in the peak flow of the rivers, typically during snow melt. The reason is the wolves' predation of the deer, and the deer's reluctance to stay in the steep-sided valleys and gullies where they are more likely to get eaten. This has allowed more trees to grow so more water soaks into the soil, reducing runoff, whilst the increase in tree numbers has in turn encouraged the beaver, whose dams further slow down the water.

The building of leaky dams made from branches is also being trialled in Britain, without the aid of beavers, to protect the flood-prone town of Pickering. This trial, and the Pontbren 'discovery' come at an appropriate time, because in recent years the weather has become more extreme and unpredictable, with heavier rainstorms causing flooding to farmland, houses and businesses, whilst bridges which in some cases have withstood natural forces for hundreds of years, have been swept away. To tackle the problem, the authorities' go-to policy is to use concrete and steel to build up embankments and barricades to get the water to flow away as fast as possible, which in turn causes greater flooding downstream, an approach mirrored by many farmers who keep their ditches clear but fail to maintain the porosity and water-holding capacity of their soils. But it is no longer working. The flooding situation in Britain now, as with the weather, is becoming more like parts of America, and the conventional 'wisdom' of water management is proving to be as ineffective, destructive and foolhardy as it was in the United States seventy years ago. -

> " ..taxpayers, property owners and insurance companies have spent billions of dollars in repairing damages and in constructing, maintaining and repairing vast dams and levees at the mouths or halfway down our great rivers, when all the time it must have been easily evident even to a kindergarten child with a pile of sand and a watering can that one does not stop floods at the bottoms of rivers but high up on their tributaries and in the forests and cow pastures." [2]

These lines were written by the American writer and farmer, Louis Bromfield, whose farm-scale experiments he recounted in his Malabar Farm books. After living abroad for two decades, much of the time in France, in 1939 he returned to his native Ohio with the intention to farm. Part of his plan was to reclaim and rejuvenate land considered to be farmed out after three generations of careless exploitative farming. The cornerstone of his approach was to use legumes, lots of legumes, mostly alfalfa, to increase the stability, fertility and productivity of the soil. As the soil improved, the yields increased, and it also allowed water which once would have run downhill, carrying some of the soil and nutrients with it, to soak into the soil and percolate downwards. After a few years, springs which had stopped running in summer started to flow all year round, providing a constant supply of fresh water for the livestock, and in so doing, demonstrated how good farming does not just benefit the farmer whose land it is, but also everyone in the locality who benefits from refilled aquifers and an ameliorated local climate facilitated by the availability of moisture during the summer.

Another enlightening experiment took place at Hof Hollergraben, [3] a biodynamic farm in Schleswig Holstein, Germany, where researchers carried out trials on two adjacent plots growing vegetables. Having been under biodynamic management for several years, the soil at Hof Hollergraben will have already been in good heart, with a good organic matter content. Both plots grew the same crop each year, and all the activities, the planting, the cultivations and harvesting, all happened on the same day on both plots, using the same machinery. The only difference was the motive power; on one plot a pair or a single horse was used depending on the job in hand, whilst the other plot was worked by a tractor. Yields were consistently higher on the horse-powered plot and after six years under this regime, the soil of the horse-drawn plot was found to have 40% greater pore space than the tractor-powered plot. Many horse farmers will not be surprised at this difference, though being able to put a figure to it adds weight and credibility to anecdotal evidence. What is most remarkable about the results is that they were obtained using a tractor which weighed 1.5 tonnes, only a fraction of the size of the tractors commonly used nowadays, and less than the combined weight of the two horses. One can only imagine how much greater the difference would have been had a larger tractor been used.

The implications of these results are many. Most obviously, the greater pore space on the horse-powered plot will make cultivations easier, requiring less power and therefore using less energy, whilst the plant roots will penetrate the soil with greater ease, as will the air, so the soil will warm up more quickly in spring. Such an open soil will also drain more quickly in normally wet times, and when extraordinarily wet weather does come, the soil has a greater water holding capacity. That means that there will be less water running off the field, into the ditches and rivers and less likelihood of water flooding people's homes.

But there is another aspect to the compaction story, and that is its relation to the production of nitrous oxide. The significance of N_2O is the effect it has on the climate, because it is nearly 300 times as potent as carbon dioxide, and as such, it is the most powerful greenhouse gas produced by agriculture. Nitrous oxide is produced by soil micro-organisms as a normal part of the nitrogen cycle, but it also comes from manures, liquid slurry being the worst culprit in this regard, and from artificial nitrogen fertiliser, especially when it is applied to excess. However, the amount of nitrous oxide released into the air is not just dependant on how much nitrogen is there, but also on the condition of the soil, more being released in anaerobic, and in waterlogged conditions, the very opposite conditions to that found on the horse-powered plot at Hof Hollergraben.

What these individual experiments don't tell you is how much working horses would affect a change in the structure of soils under different cropping, or to what extent the open structure of a humus-rich soil decreases nitrous oxide emissions. Neither do they tell you how much flooding or drought could be prevented in my soil conditions or in your climate, nor in turn how much it might translate onto the farmer's or a nation's bottom line. There is, as Geoff Morton would have said, more about it than we might think. Still, given the connections and similarities between experiences from different parts of the world, and the cumulative effect of positive measures as they impact upon each other, I can't help thinking that if we all farmed with horses, we wouldn't feel the need to send men to the moon.

Chapter Nine
The Pull of the Horse-drawn Plough

Heavy Horse World, 2020

When I was learning to plough, a piece of information which stuck with me was that the plough I was using, a Ransomes LCP, pulled 'half a horse lighter' than a normal plough. The LCP is a semi-digger, which means that the mouldboard is shorter and steeper than a general-purpose plough, and because it is also concave, it crumbles the soil as it turns, leaving it in a friable condition requiring only minimal working before it can be drilled. What Geoff Morton meant by 'half a horse lighter' was, that based on a two-horse team, the LCP turning the same size furrow only required 1 ½ horses to pull it, though in the absence of half a horse, you generally set the plough to take a bigger furrow! From experience, I knew that Geoff's knowledge was extensive and accurate, but it struck me as strange that a plough which also partially crushed the furrow as well as turning it over, in effect doing two jobs at once, would be lighter of draught than one that just turned the furrow over. So, I stored that statement as an open question in the back of my mind, and there it stayed for twenty years until I had the time and opportunity to do some research.

The Draught of the Plough

For centuries, the plough has been the primary tool for tilling the land, but during the 19th century in particular, there were major improvements. These were a result of technological changes in manufacturing, the enclosure of land from the medieval open field systems into individual farms which allowed more efficient use of the land, and the quest for advancement in all human endeavours based upon reason rather than the strictures emanating from the pulpit. Many of the early improvements were made by trial and error, but by the middle of the century the development of the draught-gauge or dynamometer, a device based around a stretched spring which indicated the force applied to it, allowed experimenters to quantify the forces involved.

Compared with other farming activities, ploughing is hard work and takes a long time, so making ploughing easier was important both economically and for the horses. Unfortunately, many of the results of the trials have not survived, including the experiments done by the plough makers themselves. Some experiments, however, were published, amongst them a trial conducted by Philip Pusey MP in 1840.[1]

At that time, there was a wide variety of different practises when ploughing; in some places three or four horses worked at length in the furrow, driven by a horseman walking beside the horses with a ploughman steering the plough, whilst others did the same work with a pair and one man. In Scotland, the swing plough (a plough without wheels) held sway, but in England most of the ploughs had wheels. Given this situation, Pusey was just as interested in finding out the best way of using the horses as in the draught of the ploughs. Being an early

Four horses ploughing in line, with two men in attendance. This picture was probably taken on the clay land in the Weald of Sussex. To keep the chains of the horses at a better angle, the middle two horses have short chains into the main traces, and the wheeler has a cart saddle to carry the downward pressure created in the traces of the other three horses.

experiment, his trials were somewhat limited by circumstance and by the inadequacies of the draught gauge, which Pusey suspected was not particularly accurate under very light loads. Nonetheless, his results do show some of the essential differences between ploughs, which continued to influence their design for years to come.

The trial was undertaken in Berkshire on different soil types, using nine ploughs that were either available locally or were gaining favour for various reasons, but for the sake of clarity, I have only included six. -

- A traditional Berkshire plough with a high gallows, (a frame with two wheels which supported the front of the plough beam) made completely of wood.
- An improved Berkshire plough with one wheel and an iron breast, made by a Mr Hart from Wantage.
- A Ransomes 'Rutland' plough, previously shown to be light of draught.
- A Ransomes 'FF' plough delivered especially for the trial, smaller than the 'Rutland', and having two low wheels, originally designed for market gardening with a single horse but also used with a pair on light land in Cambridgeshire.
- Another Ransomes 'FF' plough, but without wheels.
- An improved Scottish swing plough made by a Mr Ferguson near Stirling.

Mark Morton's bodkin hitch at the Festival of the Plough at Epworth in 1992. The rear pair are yoked in the usual fashion, except the draught chain to the plough is attached to an evener one third of the way along to give the land horse equal purchase against the two horses in the furrow.

Before comparing these results, it is useful to know how much a horse can reasonably pull. Although a horse can move much greater loads in short bursts, it is widely accepted that over a working day a fit horse can exert a force equivalent to one tenth of his weight. So, a good-sized cart horse weighing 17 cwt *(which is 136 stones, 1904 lbs, or 863kg)* can pull 13.6 stones *(0.85 kN)* when measured as a force on the draught chain.

Applying this information to the chart, it is clear that the best ploughs on the lightest land would be fine for a single horse, but on the heavier land, although the better ploughs fell within the 10% guidelines when drawn by a pair, many required more power. Looking in more detail, on lighter land, (trials 1 and 3) the worst plough required 75 % more power than the best plough, and the swing ploughs required 25% more power than the wheeled ploughs. Particularly interesting is the difference between the two otherwise identical Ransome 'FF' ploughs, where the swing version required at least 20% more power.

	Trial 1	Trial 3	Trial 4	Trial 5	Trial 6
	Sandy loam	Loamy sand	Strong clay loam	Clay over-lying Kim-meridge clay	Poor moory (peat) soil
Old Berkshire	23	21	36	32	25
Hart's improved Berk-shire	14	12	23	28 wheels sunk	16
Ransome's 'Rutland'	-	16	36	-	21
Ransome's FF wheeled	14	13	33	wouldn't work	14
Ransome's FF swing	18	16	30		21
Fergusan's improved Scottish swing	19	19	35	24 (31 when repeated on troddden ground)	23

Pusey's trials of different ploughs on different soils, all ploughing a furrow 7" wide and 5" deep (17.5 cm x 12.5 cm). The draught of the ploughs is given in stones of force.

On the heavier loam soil in trial four, however, the difference between ploughs was much less, all requiring 30 to 36 stones of draught *(1.87 to 2.24 kN)*, apart from Mr Hart's plough which was still significantly easier to pull. On the true clay ground in trial 5, however, the surface was so soft that the wheel of Hart's plough was pulled through the mud as the plough sunk in, but not being a ploughman himself, Mr Pusey did not think of getting the ploughman to try the plough with the wheel removed. In this soil, the Ransomes wheeled 'FF' plough could not be got to go at all, presumably because of clogged or sunken wheels.

In the fifth trial, Pusey experimented with different hitches, the old Berkshire plough being pulled by four horses in line as was traditional in the area, whereas the Ferguson plough was used with a pair yoked abreast. Although the results showed the Ferguson to be much easier to pull, the land horse struggled on the very wet ground, a fact not lost on the local farmers who suggested that the Ferguson be tried again after the ground had been further cut up and trodden by horses pulling three other ploughs. On this second run, the draught of the Ferguson increased to 31 stones *(1.9 kN)*, very similar to the old Berkshire, a warning to anyone that the condition of the soil is of primary importance when ploughing.

These results conform with the agricultural practice of the late nineteenth and early twentieth centuries, where most land was ploughed with an abreast team and a wheeled plough, but heavy land was ploughed without wheels, especially when wet, either with the horses walking behind one other in the furrow, or yoked bodkin fashion, where one horse is on the land, and two walk in the furrow.

In the third trial plot, two more experiments were made. The first compared ploughing at different depths. Until then, and I am sure some people still believe it now, it was widely held that the draught of the plough is proportional to the depth of ploughing, but these results

width of furrow	4 inches *(10 cm)*	5 inches *(12.5 cm)*	6 inches *(15 cm)*	7 inches *(17.5 cm)*
Old Berkshire	21	21	24	31
Hart's improved Berkshire	11	12	16	18
Ransome's Rutland	16	16	18	20
Ransome's FF wheeled	12	13	18	21
Ransome's FF swing	16	16	18	21
Fergusan's improved swing	18	19	19	22

Pusey's experiment to compare the draught of various model of plough ploughing at different depths, all cutting a furrow nine inches (23 cm) wide. Figures are in stones of force.

clearly show that not to be the case, the average draught of all the ploughs only increasing about 20% as the depth increased by 50%, from 4" to 6" *(10 to 15 cm)*. Indeed, except for the old Berkshire plough, when ploughing at 6 or 7 inches *(15 to 18 cm)*, all the ploughs were very similar in draught.

Yet another trial on plot three measured the force required to drag the ploughs along the surface of the soil. Here the Ferguson swing plough registered 12 stones in draught compared to its estimated weight of 15 stone, comparable figures for the Ransome 'FF' swing plough being 10 stones to 10 ½, and 3 stones of draught to 8 stone of weight for Hart's plough. From our point of view today, these figures should be treated with caution, as one plough might have been running on the wheel and the heel of the sole whereas another might have had the whole length of the sole in contact with the ground. Nonetheless, it shows that the friction between the plough and the ground is a significant part of a plough's draught.

The last of the experiments compared ploughing at different speeds, Pusey being impressed by the work rate of a pair of Clydesdales lent by his friend, Lord Moreton. Contrary to expectations, the plough's draught increased by only 4% between a very slow walk of 1 ½ mph *(2.4 km/h)* and a very fast walk of 3 ½ mph *(5.6 km/h)*. However, we must go to another study, described in J C Morton's 'A Cyclodedia of Agriculture,' [2] to discover that although the average draught only increases slightly, the oscillations in the draught are doubled by doubling the speed. Since it is the peaks of draught, where the plough meets high resistance such as a stone, which is most jarring for the horses, ploughing at such high speed is not a good solution. In addition, although the plough may only require a little more effort to pull it, the horses also need to move themselves faster which takes considerably more energy, as any walker who starts running will be able to tell you.

Mr Ransome's Lecture

In 1865, twenty-five years after the Pusey trials, the Royal Agricultural College invited James Ransome of the Ipswich-based implement makers, to talk about ploughs and ploughing. [3] Quoting his own experiments to demonstrate the variations in draught, he confirmed Pusey's findings on wheels. By measuring the draught of two otherwise identical ploughs, one with wheels and one without, he showed that despite the extra weight of the wheels, it recorded 22 stone of draught *(1.37 kN)* compared with 30 stones *(1.87 kN)* for the swing plough, though he also pointed out that this benefit is lost once the wheels get clogged with soil.

He then illustrated the effect of the plough's weight, his company, amongst others, trying to maintain strength whilst reducing the weight of their ploughs. In another experiment he added weights to a 24 stone *(155 kg)* plough, adding 8 stones *(54 kg)* each time. The plough on its own recorded a draught of 21.5 stones *(1.34 kN)*, with an extra 8 stone *(54 kg)* of weight, the draught increased to 28 stones *(1.74 kN)*, and with 16 stone *(108 kg)* added, the draught was 33.5 stones *(2.08 kN)*.

Before finishing his comments on plough draught, Mr Ransome stressed the importance of a thin, sharp coulter and share, because a large part of a plough's work is in cutting the furrow slice, though he did not say how much that might be. He also compared the plough's action to a wedge, the thinner the share, the more easily it will work.

At that time, the mouldboards of most English ploughs were relatively long, what we now call a general-purpose mouldboard. Within a few years, however, a new type of plough with a short mouldboard was being imported from America amidst claims that this plough was easier to pull. This novelty was the Chilled Plough.

The Chilled Plough

The first chilled ploughs reportedly came to Britain in 1879 and looked very different from British ploughs, being heavier in build, usually with a wooden beam and with one wheel. Made by the Oliver factory in South Bend, Indiana, the imports increased substantially when Wallace of Glasgow became their sole agents in 1885. The big difference with these ploughs, and the reason they got their name, was because the mouldboard was chilled during manufacture. This was achieved when casting the molten metal, the peculiarity of the process being that instead of the metal being poured into a mould completely filled with sand, the shape of the wearing surface of the new mouldboard was defined by a piece of iron which induced rapid cooling in the molten metal, changing the crystalline structure of the outer surface and making it very hard. The process of chilling was already used in Britain for shares, James Ransome's grandfather, Robert, having gained the patent for a chilled share in 1803, but this technique had not been applied to mouldboards. The other difference with the Oliver mouldboards was that they were concave, or hollow from top to bottom, so the furrow slice was crushed together as it was pushed against the mouldboard, making the furrow slice crumble. This contrasted with the generally flat English mouldboards and the convex mouldboards of the Scottish swing ploughs, the furrow slice in both cases maintaining its shape as it turned.

Concerning the draught of the Oliver plough, in a paper given to the Airth Agricultural Class in 1892,[4] John Speir gave figures based on his own trials. When ploughing a 13-inch *(33 cm)* furrow, the Oliver plough required a force of 16 stones *(1 kN)* to move it, whereas a Scottish swing plough taking a narrower furrow of 10 ½" *(27 cm)* required nearly twice as much, 30 stones *(1.87 kN)*, both ploughs working at a depth of seven inches *(18 cm)*. After repeated tests over three years with different swing ploughs, the American plough consistently performed better, Spier attributing most of the improvement to the hardness of the chilled mouldboard. However, given the higher draught of the Scottish ploughs in the Pusey trials, in part because of the lack of wheels, the comparison might not have been quite so favourable if a wheeled plough had been used as a comparison.

Nonetheless, the new chilled plough was a major improvement, and British manufacturers soon began producing their own versions. Although some farmers were reluctant to adopt them,

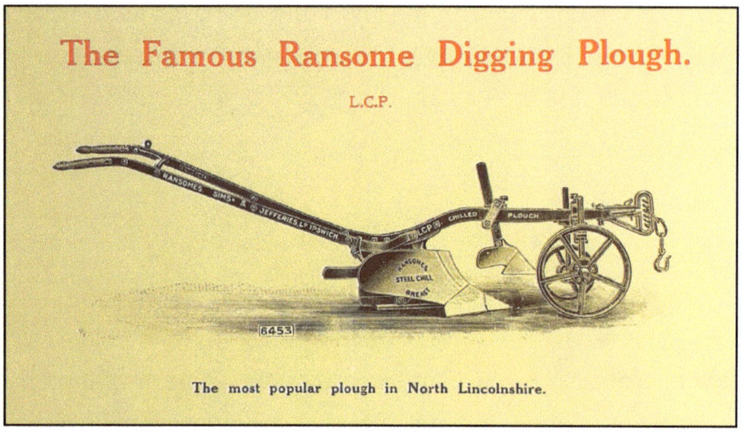

the Standard Cyclopedia of Agriculture [5] noted, “they have proved to be the most economical ploughs to use under most conditions,” going on to say, “repeated tests have shown that in ploughing firm land the draught of the modern chill breast digging plough is about one third less than that of an ordinary plough.”

One third less is not so different from Geoff Morton’s figure of ‘half a horse lighter,’ but although these figures are a good guide under most conditions, the use of general-purpose mouldboards continued. There are probably three reasons for this; the conservatism of farmers played its part, as did the continuing belief, though disproved, that you needed to have a crested furrow to allow the frost to break it down, when in fact normal frosts would (in the past) penetrate two inches *(5 cm)* into the ground, regardless of the shape of the furrow slice.

The third reason, however, has more substance, and has to do with the makeup of the soil. The critical factor is whether the soil will crumble under the action of the concave mouldboard. If it will crumble, the draught is less than when using a general-purpose mouldboard, but some soils with a large proportion of clay are so tenacious that they stay together in one piece, in which case the digging plough cannot crumble the furrow slice. In this case, turning the furrow intact with a general-purpose mouldboard takes less power.

For us nowadays, there are some additional factors to consider. Compared with fifty or a hundred years ago, many soils are significantly depleted in organic matter. Organic matter or humus does several things in soils; in sandy soils it helps to stick it together and to hold more moisture and nutrients, whilst clay soils become freer draining and more friable. In all soils, organic matter encourages more soil life, such as earthworms which help keep the soil open and more crumbly. Fields which were once rich in humus and easy to plough will therefore be harder to plough and will tend to crumble less readily than years ago. Another issue is soil compaction caused by heavy machinery, by cultivating when too wet and compounded by a lack of humus. Soil compaction is one reason why ever bigger tractors are needed to pull the same size ploughs, and why getting a horse plough to work can be harder than it should be.

A view of the LCP plough, the straight electric fence post highlighting the hollow in the mouldboard, which crumbles the furrow slice in on itself as the flat underside is forced to conform to the mouldboard's shape.

Modern Mouldboards

Whilst those who plough with horses in Britain tend to use ploughs which are at least seventy years old, in North America, new ploughs are being built, principally by two Amish companies, White Horse Machine and Pioneer Equipment.* Given the technical challenges of manufacturing mouldboards, these firms started out by using the wearing parts, the shares, slades and

* *Pioneer no longer make horse-drawn equipment, though some of their designs are made by others.*

mouldboards, made by Oliver and John Deere for their tractor ploughs. These mouldboards are of the sharper-turning digger type, more severe the the LCP mouldboard, which work well on lighter land, but on heavier land horse farmers were finding that the furrow tended to buckle as it was thrown over, leaving an uneven finish and adding to the plough's draught. So these companies now offer the option of a slower turning mouldboard, made by the Norwegian company Kverneland, or a similar one by Keystone. Initially there were reports of the Kverneland being significantly lighter of draught, but tests made by Pioneer show that it depends on the soil, "the Oliver and John Deere bottom performing better in loose or sandy soil, especially when plowing corn (maize) stalks or oat stubble." [6]

A modern reversible plough made by White Horse Machine.

This contrasts with grassland, when the Kverneland and Keystone work better, partly because of the vigorous root systems of the newer grass and clover varieties which hold the soil together more. For practical farmers, however, Pioneer point out that the effective inversion of the soil for better weed control is more important than merely focussing on draught, though for shallow ploughing, the Kverneland is more successful than the Oliver at burying weeds.

Friction and the Rear Wheel

As shown in Pusey's early experiments, friction is a major constituent of a plough's draught. Some friction is unavoidable, between the soil, the share and mouldboard as the furrow slice is turned over, but there is also a considerable amount of friction on the slade. This is the part of the plough that runs in the corner of the furrow and keeps the plough running straight and level, functioning like the runner of a sledge, distributing the pressure created by the furrow slice pressing on the share and mouldboard. Aware of the power required to overcome this friction, plough makers in the early nineteenth century experimented with reducing the size of the slade or replacing it with a wheel behind the body of a plough. The earliest 'friction wheels,' as they were then called, didn't work well on stony or soft ground and the relatively primitive bearings tended to wear too quickly, making the plough run unsteadily. However, improvements in the late 19th century, such as easily replaceable bearings, made friction wheels more successful and they become more common, particularly on two-furrow ploughs.

Mr Ransome on Two-furrow Ploughs

In 1872, James Ransome gave another talk, this time on the subject of two-furrow ploughs. [7] His upbeat assessment focussed on the advantages of two-furrow ploughs over single-furrow ploughs, and the benefit of friction wheels. This is how he summed up the variables on the ploughs' draught;

- The old type of double furrow ploughs which had two slades, one on each body, took twice the power of a single plough fitted with a slade.
- A double furrow plough, of which the back part of the front body is cut away and which only has one slade on the hind plough, does not take more power than 1 ¾ times of a single plough with a slade.
- A double plough of the new sort, with a friction wheel behind instead of a slade, does not take more than 1 ½ times the power of a single plough with a slade.

However, Mr Ransome did qualify these remarks by saying that on hard dry ground there is little advantage in having a friction wheel, but they were a very advantageous in wet conditions, especially on sticky soils, when the suction between soil and slade was considerable. The implications of these results were clear; the farmer could save time, labour and horses, and keep on top of the work by switching to double furrow ploughs.

When Ransome gave this talk, the chilled plough was yet to arrive, but combining the benefits of two furrows with chilled digging mouldboards, one of his main competitors, Hornsby and Sons of Grantham, had this to say in their catalogue of 1891; [8]

- The advantages of using a double-furrow plough are even more striking in the case of diggers than in that of general-purpose ploughs.
- Two ordinary (general purpose) ploughs will each turn one furrow 9 inches *(23 cm)* wide, 18 inches *(46 cm)* for both ploughs, requiring four horses and two men; the work thus done by both ploughs in a working day averaging 2 acres *(0.8 ha)*.
- One double-furrow digging plough will turn two furrows each 12 inches *(30 cm)* wide, 24 inches *(60cm)* in one operation, only requiring three horses and one man; the work thus done in a working day averaging 2 ½ acres *(1 ha)*.

At that time, Hornsbys did not offer a rear furrow wheel on their two-furrow ploughs for the domestic market, had they done so, the daily output of the double-furrow digger could easily have been over three acres. Their ploughs made for export, however, whether two-furrow walking ploughs or the ride-on gang ploughs, did have a rear furrow wheel, Hornsbys describing the elimination of the slade in favour of the rear furrow wheel as "giving extreme lightness, both in draught and weight."

Ride-on Ploughs

The ride-on plough was an American invention, the single-furrow 'sulky' plough and the two- or three-furrow gang plough gaining popularity from the late 19th century, though they were exceedingly rare in Britain. These ploughs were heavier than the walking ploughs, especially the ones with a separate frame from which the plough bodies are suspended (known as a high-lift plough) but the saving of draught by having cut-away bodies, digging mouldboards and a rear furrow wheel easily counteracted the extra weight of the plough frame and driver.

Three abreast on a double furrow plough, very probably in north Lincolnshire. In these open spaces, it is easy to see the advantages of these two furrow ploughs, with one man being able to do twice as much work and the three horses working no harder than if two of them had been used on a single furrow plough. Nonetheless, single furrow ploughs remained much more common than two-furrow models.

Two six-horse teams of Percherons with two-furrow gang ploughs, ploughing the huge fields of North Dakota in the 1920s or 30s.

The use of digging ploughs and riding ploughs with a rear furrow wheel contributed towards the work rate of American ploughing. Whereas an acre *(0.8 ha)* a day was the standard in Britain for a day's ploughing, although in fact, on heavy clay ground that would be good going, and on light land up to two acres *(1.6 ha)* was achievable, in America they aimed at ploughing an acre *(0.8 ha)* per horse per day. [9]

Another reason for this higher output was that the American ploughman did not need to hold his horses to a slower place when riding on a gang plough to make his own life easier, whereas that was a common complaint from British farmers when talking about their own trudging ploughmen. But regardless of opinion about the merits of ploughmen, if you can do a job where the horses are walking freely, that is when they cover most ground and when more work is done, so if you can balance the number of horses with the draught of the implement to change heavy work to moderate work, that is all to the good. In many parts of America they also had plenty of horses, so adding an extra horse, meaning that each horse in the team was working less hard, was a more realistic proposition than it was in Britain. This is one reason why many US horsemen like to use four horses, yoked two and two, on a sulky plough that can be handled by three abreast. But there is another reason which is of fundamental importance to how hard a plough is to pull, and that is minimising what is known as side draught.

Pioneer Equipment's foot-lift plough. The wheels and seat are fixed to a frame, while the plough body is suspended underneath, being put into work and lifted at the end of the furrow with foot pedals.

Side Draught

Side draught is the sideways force that is applied to the plough, mostly on the slade, when the horses are not hitched in line with the centre of resistance of the plough. Side draught is not only wasted energy because it does not contribute to moving the plough forwards, but it also increases the friction and therefore the draught of the plough. It is like pushing a car, but instead of pushing at the back in the direction of movement when all your effort goes into moving the car, you instead push diagonally on the wing, the sideways part of the force only serving to increase the friction of the tyres against the ground.

With a single-furrow plough, the centre of resistance, usually known as the line of draught, is nearly under the plough beam. If you yoke your horses so the draught chain to the horses' eveners coincides with the line of draught, the plough runs straight without significant sideways force, and all the force applied by the horses is effectively transferred to pulling the plough. However,

The rear furrow wheel of Geoff Morton's gang plough counteracts the thrust of the furrow slices as they pass over the shares and mouldboards. The angle of the wheel indicates the proportion of the force that is trying to push the plough sideways.

once the draught chain from the horses deviates from the line of draught, the horses' efforts apply a turning force on the plough, which, because the plough can't turn because its design keeps it in the furrow, expresses itself as greater pressure on the slade and landside.

To put that into real terms (see illustration below), with a twelve-inch furrow, the line of draught is about 1 ½ inches *(4 cm)* away from the vertical cut of the plough, so therefore 10 ½ inches *(27 cm)* from the pre-existing furrow wall. Because the furrow horse walks down the middle of the open furrow, the centre of his swingletree is also in the middle of that furrow, 6

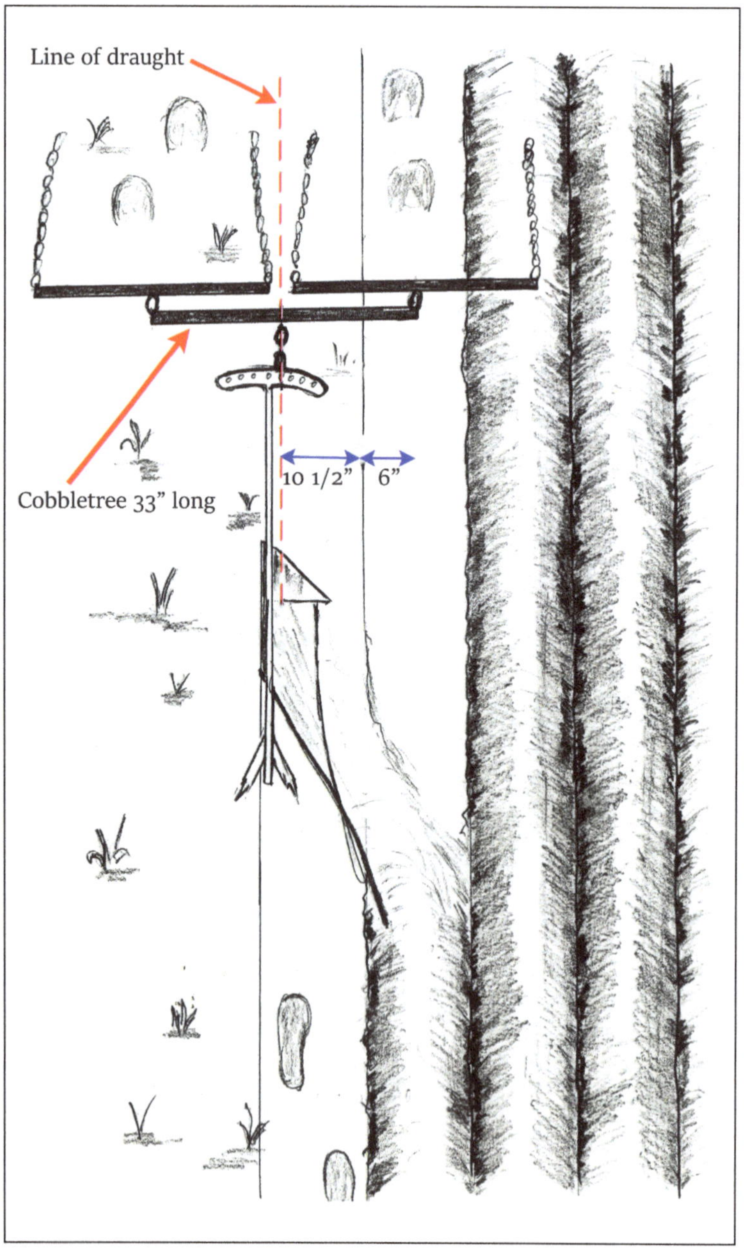

inches *(15 cm)* from the furrow wall and therefore 16 ½ inches *(42 cm)* from the line of draught. For the land horse's efforts to balance the furrow horse, the cobbletree should therefore be twice as long, 33 inches (84 cm).

However, in Britain many ploughs cut a much narrower furrow, so for eight-inch furrow *(20 cm)*, the ideal length of the cobbletree should theoretically be 22 inches *(56 cm)*. This is too short to allow for swingletrees long enough to keep the traces away from the horses' sides and legs, so with narrow furrows it is not possible to hitch two horses abreast along the correct line of draught, so inevitably there is some energy which goes to waste due to side draught. This is probably why the line of draught is never mentioned in British literature on ploughing. However, the nearer you get to yoking the horses in the optimal position, the easier the plough will run and the easier it will be for the horses.

A pair of horses yoked to a plough cutting a nine inch furrow, although this model is capable of ploughing up to twelve inches. To keep the plough running true, the hake is offset to the left, inevitably creating some side draught.

To give an idea of the importance of side draught, in a 1928 publication by the Horse and Mule Association of America,[10] they quoted the results of a test using a dynamometer to compare two different hitches on a two-furrow gang plough. The results showed that four horses yoked abreast, with one in the furrow and three on the land, had to work between 15 and 25% harder than when the horses were yoked two and two, along the line of draught.

The Furrow Slice

Under most conditions, a plough turns a furrow slice first by 90 degrees, onto its edge, and continues to push it over another 45 degrees so that each side of the final furrow slice is of equal width and equally exposed to the elements. During the first 90 degrees of turn, the mouldboard is lifting the furrow slice, so the wider the furrow is, the higher the furrow slice must be lifted. Pushing against gravity, this takes more power, and, because the amount of cutting and shearing done by the share is also proportional to the width, it should be no surprise that the draught of the plough is also largely proportional to the width of furrow.

Increasing the depth of the furrow, however, does not necessarily increase the draught proportionally, some of the ploughs in Pusey's experiments measuring the same draught when working at five inches as four inches *(12.5 compared with 10 cm)*. At six inches *(15 cm)* however, the draught of most ploughs increased significantly, but there was a great variation between the models, showing how some designs are more suitable for ploughing deeply and some are better when ploughing more shallowly. Deciding on the depth of ploughing is dependent on several factors, how deeply it was ploughed last time, the quality of the soil and whether you want, for instance, to go a little deeper so the plough sits more happily in the ground. The proportions of the furrow slice also affect the running of the plough, the ideal proportions being 7 to 10. Even if you are only ploughing a furrow eight inches *(20 cm)* wide, this proportion still means you are ploughing to a depth of 5 ½ inches *(14 cm)*, which may be more than you need, either to bury the weeds, or if you want to minimise the mixing of the soil layers and upsetting the soil bacteria. Also, at this depth you are starting to ramp up the power requirements, making more work for the horses.

The 7 to 10 ratio becomes more problematic with modern horse ploughs which use big plough bodies designed for tractors. Because each body is wide, they tend to work best at greater depth too. To put that into figures, a twelve-inch *(30.5 cm)* furrow should theoretically be working nearly 8 ½ inches *(20 cm)* deep, and by the time you get to a sixteen-inch *(40 cm)* furrow, it should be ploughing eleven inches *(28 cm)* deep, which is obviously excessive, both for the soil and for the exertions of the horses. In practise, you can run ploughs shallower than the theoretical ideal, but there is a point to be made about excessive energy use due to over-deep ploughing, whether by tractor or horse. Although not available at the moment, in terms of power requirements and soil preservation, there would be a lot of sense in using a three-furrow plough with 9-inch *(23 cm)* bodies rather than a similarly wide two-furrow plough with 14-inch *(36 cm)* bodies.

Coulters and Skimmers

Of the work done by a plough, nearly half can be attributed to cutting the furrow slice. This is why a sharp share makes so much difference to the plough's draught. The other half of the cut, the vertical cut, is made by the coulter. The knife coulter, which was the usual type on British ploughs, is about half an inch *(1.2cm)* thick and has a blunt bevel on the leading edge, which makes it strong enough to break through roots and dislodge stones, but its thickness and bluntness mean that it is harder to pull through the soil than a sharp, thin, rotating disc coulter. The angle of the bevel also means that the knife coulter pulls the plough more towards the land, so in most instances this will add to the side draught. The disc coulter comes into its own when there is a big mass of roots to cut through or when there is a lot of surface vegetation. On digger ploughs with a shin on the leading vertical edge of the mouldboard, there is no need to use a coulter, though a disc coulter would reduce the draught.

The skim coulter, in ploughing its own little furrow, also adds to the plough's draught. Depending on what you are ploughing and why, you can dispense with the skimmer entirely and save some draught, but if you want to bury all the vegetation it is a necessary addition. Many ploughmen are very particular about setting the skimmer, but as far as draught is concerned, the less material the skimmer removes, the less power is required. The first requirement, however, is that the skimmer is set deeply enough to do an effective job.

Draught Springs

One of the most important variables on the plough horses' workload is the fluctuation in draught. This changes considerably from second to second, the peaks of draught, when the plough encounters a stone or more solid piece of ground, for example, being considerably more than the average draught. If you can lower these peaks of resistance, you minimise the jarring on the horses so they are less tired by the end of the day. To achieve this, in the nineteenth century draught springs were invented to cushion the load, and modern versions are used by some farmers and horse loggers today. Some modern ploughs also have a spring incorporated into the hake to minimise the shock. As far as I am aware, there is no consensus or even discussion about whether draught springs or elastic traces should remove only the very worst of the shocks but do little or nothing to even out the normal fluctuations, or conversely, be weak enough to even out most of the variations, except for the major shocks. However, the presence of spring-loaded hakes shows the manufacturers do take it seriously, and for anyone doing a lot of ploughing, especially on heavy or stony ground, draught springs are worth serious consideration.

Spring-loaded hake on White Horse Machine's walking plough.

Reading the old literature, it is clear that our predecessors were highly knowledgeable about the design and use of horse-drawn ploughs, but particularly with design, much of that knowledge has been lost because nearly a century of abundant and cheap fossil fuels has meant that plough draught has been of little interest. In recent years however, developments by companies such as White Horse Machine and Pioneer who understand the mechanics of horse work, in conjunction with mainstream plough manufacturers such as Kverneland, have advanced the design of horse ploughs significantly. We may be some way off having the variety of ploughs once available or a widespread understanding of plough draught, but at least we are at least moving in the right direction.

Chapter Ten

Lancaster County

This is a translation of an article I wrote for the German draught horse magazine, Starke Pferde, combined with another that appeared in the Small Farmer's Journal in 2011.

Just before the fourth of July weekend, I was lucky enough to be part of a group travelling to the gently rolling landscape of Lancaster County, Pennsylvania, home to the oldest Amish community in the United States. Our destination was Horse Progress Days, an event I had long been wanting to visit, the biggest show of real horse power and new horse-drawn equipment in the world. We arrived a couple of days before the event, time to acclimatise to the time zone and the humidity, and to call in on some of the local companies who make horse-drawn machinery.

On the first morning I awoke much too early, my body clock still on Shropshire time, so went for a long walk before breakfast, along the rural roads and in amongst the farms. In the cool of the early morning there were few people about, but in the lowest level of the barns the cows were already being milked, whilst outside workhorses and mules lazed behind outbuildings and under the shade of trees. The farms of Lancaster County are a delight, with orderly farmsteads, well-tended vegetable gardens and soils in good heart, a testament to generations of good stewardship. Even the farmhouses bear witness to this continuity, many having been extended to the sides, often more than once, to accommodate the next generation of farmers or the grandparents as they became less active.

Mules grazing in the cool of the early morning. In America, especially in hot climes, draught mules are often used in preference to horses.

By the time I returned for breakfast, the horses and mules were out in the fields, working down the ground after the cereal crops had been harvested, their steady pace marking a huge contrast to the trucks hurtling along the Lincoln Highway. Later, as we drove to our first visit, there were horses and mules at every turn, pulling discs or cultivators, sprayers and mowers, prompting one of our party to remark, “oh, there’s someone in a field working horses, and it’s not me!”

An Amish girl uses a pair of mules and a well-used hitch cart to roll the field.

On the day before the event, we joined a coach tour round several more Amish businesses, a company making steel wheels for wagons and other farm implements, a workshop producing biothane harness, a typical Amish dairy farm and another making products from goats' milk. Between each stop, the coach passed many more well-managed and beautiful farms, with draught horses and mules working the fields and horse-drawn buggies trotting along the roads.

New hitch carts lined up outside the workshop of White Horse Machine. The avoidance of pneumatic tyred vehicles amongst Old Order Amish groups means steel wheels are used on farm machines. The handle on this model of hitch cart allows the wheels to be steered, useful when backing up precisely to hitch onto an implement, and to avoid an implement crabbing downhill on side slopes.

For those of us who work horses in relative isolation, seeing a whole community involved in animal-powered agriculture is quite an experience, so as the coach dropped us off at the end of the day, our excitement at going to the show only increased, as in the distance, the rows of new horse-drawn equipment became longer and longer.

It was therefore with no little anticipation that I descended the steep grassy slope from the car park the following morning, overlooking rows of buggies lined up next to the long tent used as a stable, and beyond that, marquees full of trade stands, lines of equipment, and everywhere horses, mules, and crowds of people converging on the field demonstrations. Coming to the Horse Progress Days for the first time, the scale and variety is scarcely imaginable; although I had expected to see well-engineered, functional equipment and well-behaved horses and mules, I had not imagined the sheer variety of horse-drawn tools, the number of people, or the fantastic collaborative atmosphere that underpins the whole event, the shared interest in working horses bridging any cultural or language barriers. So even though I was thousands of miles from my native soil, being in a place where others understood about the line of draught and got useful information from picking up a handful of earth, I felt right at home.

In America, there are many more people using draught horses than in Britain. As in Europe, there came a crunch point in the 'seventies when the old horse-drawn equipment became worn out, so some farmers started to build their own. At first, many copied the old machines, but soon they started to create their own designs, sometimes incorporating parts made for tractor machinery. Over the years, the quantity and range of equipment has grown, so there are now several small factories and larger workshops producing horse-drawn equipment full

Above - A new manure spreader made by E-Z spreaders, pulled by a pair of Brabant horses wearing D-ring harness.

Left - This smaller spreader has the same range of adjustments as the larger machine but is small enough for two ponies.

Right - Pioneer Equipment's Homesteader, a multi-purpose implement based on a traditional straddle-row cultivator, seen here with a set of discs, though it can also be used with a plough body, cultivator, ridger or single row drill.

time, as well as numerous farmers who run a sideline making a few implements in the winter months. It was to showcase these developments that a group of Amish farmers got together in the 'nineties to organise the first Horse Progress Days, an event that has steadily expanded, rotating through the states which have the largest Amish communities.

Above - Six horses and a Pioneer two-furrow gang plough. The high platform gives the driver good visibility over the horses, the adjustments of the plough being via long levers.

Below - A mechanical transplanter made in Michigan with a water tank to give the plants a good start.

Three mules pull a modern three-row cultivator, the fine steering of the machine being done by the driver's feet.

A common solution in Amish communities for using horses with machines that need motorised power - mounting an engine on the implement, in this case a combination of a mower with a conditioner following behind.

Two methods to do the same job. Above - a trailed mower made by I and J Manufacturing featuring ESM's double reciprocating cutter bar. The power comes from the ground-drive forecart, also made by I and J, in this case with a battery-powered hydraulic system to raise and lower the cutter bar.

Below - Four Belgians pull a New Holland haybine, which cuts and crimps the grass, the mechanism being driven by a powerful engine mounted on the hitch cart.

A ground-drive hay tedder powered by belts from the wheels, and below, an engine-powered four-rotor tedder, both set-ups being an easy pull for a pair of horses.

For frequent visitors to Horse Progress Days, much of the machinery is familiar and has become well-loved and well-used by horse farmers, but each year new designs appear, whilst the older models are continually reviewed and improved. For anyone farming with three or more horses, it seems like you can get anything that you could possibly wish for, which is not surprising given that the Amish provide about 80% of the market in horse-drawn equipment, and they tend to farm with four to seven work animals. Of course, that means that 20% of the machinery is bought by non-Amish people, and, according to two Lancaster-based manufacturers, this part of the market is growing faster than amongst the Amish.

A small set of discs for a single horse made by Groffdale.

One of the many combination implements incorporating cultivator tines with rollers or crumblers. This narrow example for a pair of horses was made by Shipsche Farm Supply.

Although the selection of tools is wide, there is also a gradual meeting of different approaches. Most of the cultivators, for example, now come with a crumbler which allows two jobs to be done in one pass without a great increase in draught, and the consolidated ground is easier for the horses to walk on during the next operation. With the hitch carts too, most manufacturers offer the choice of brakes and hillside steering, with the option of hand- or electric-powered hydraulic pumps or a small engine to drive a PTO, giving the simple hitch cart incredible versatility. Less harmonious were the monster machines requiring a a four-wheeled hitch cart with a huge engine, normally requiring at least four horses to keep the whole thing moving, an approach sometimes criticised because of the machine's size and use of fossil fuels, though within the Amish context, these machines make more sense, but there were also plenty of ground-driven machines. For those farming with a pair or just a single horse the choice is more limited, but with many Amish farms getting smaller, some concentrating on vegetable production, more implements of this size are being produced.

This New Holland baler has an engine mounted on the baler itself. The motorised baler alone would be a load for two horses, the bale collector making up the load for one or two more.

With an event as big as Horse Progress Days it is impossible to see everything, and hard to take in everything you do see. Although I mainly went to look at the machinery and to see what others do with their draught animals, I won't forget the delicious horse-powered ice cream, the tiny wagons drawn by ponies bringing more bottles of water to the ice-filled containers and, as the afternoon progressed, the groups of Amish teenagers playing volleyball in what, until earlier in the day, had been a hayfield. Nor will I forget the surprised and

Above, and at the beginning of this chapter - Six splendid draught mules pull a John Deere baler behind I and J's heavy duty ground-drive forecart. In these conditions without any sledge or bale collector behind, four would have been enough.

Detail of the gearing for the power take off shaft underneath the platform of I and J's smaller ground drive forecart.

As many farmers are reducing or eliminating ploughing in favour of minimal till or no-till cropping, this crimper/roller, made by I and J Manufacturing, is designed to knock over and break the stalks of cover crops. Seen here behind a basic hitch cart, the battery-operated hydraulics are mounted on the implement's drawbar.

An Esch no-till drill behind a steel-wheeled forecart.

welcoming faces when I introduced myself to the Mennonite family who reconditioned and shipped my mower all the way across the Atlantic, or sitting in the shade and comparing notes with an Amish man who makes manure spreaders as well as doing work with his two horses, and like us, grows a big garden and keeps three pigs each autumn for his family's consumption.

For anyone with an interest in working horses, Horse Progress Days is hugely impressive, but besides being enjoyable, relaxed and entertaining, its real purpose is to provide a gathering place for farmers, horse loggers and others who need this technology for their daily lives. For anyone who values a dynamic living agriculture, Horse Progress Days is a jewel, exemplified by people who calmly work their farms with intelligence and hard work. To experience just a small part of this way of life was a little strange, but at the same time exciting, reassuring and inspiring. Even after two weeks of being back at home, my head still hasn't completed the journey back across the Atlantic; some of it is still processing the impressions, the information and the atmosphere of some very memorable days.

Opposite - A young Amish woman cultivates the ground with four mules and three horses after a cereal crop had been harvested.

Chapter Eleven

The Rhythm of Horses in the Landscape

Small Farmer's Journal, 2014

The exciting thing about going to new places is that you never know what you will see or learn. One such occasion was the British Festival of the Working Horse, where the weather was dreadful and the range of equipment more than a little disappointing, but there was an unexpected highlight. This was Henry Finzi-Constantine's presentation to the mini conference, in which he talked about the introduction of a working horse into his Italian Biodynamic vineyard. His wholehearted enthusiasm for the wine, the vineyard and the horse was both infectious and delightful, but for me at least, also slightly embarrassing, because in describing what the horse brings to his operation, they were no caveats, no ifs or buts.

Instead, he explained why an active and carefully managed soil is fundamental to producing healthy crops and a high-quality product. From a biodynamic point of view, the farm is seen as a whole, from the soil and water quality to the plants, animals and people, but until the horse was introduced into the vineyard there had been no animals on the place. Once the horse arrived, it provided manure to fertilise the soil and power to cultivate the vineyard, with some of the work between the vines being done more quickly, and other jobs probably more slowly. But Henry's main point was nothing to do with how useful a horse might be, but simply that it belonged in that situation. His most striking observation was that the whole atmosphere in the vineyard changed with the arrival of the horse, bringing a living focus of interest and completing the circle between the plants, the soil and the people. This might seem a bit airy-fairy, but there is a fundamental truth underlying this observation, one I have witnessed on horse-powered farms from seemingly austere Amish country to dour Yorkshire.

In the 1980s, when open farms were in their heyday, Geoff Morton opened his farm during the summer months, and despite minimal advertising, we got a steady trickle of visitors. The busiest time was during harvest, which was also the busiest time on the farm. Some of the visitors just wanted to see the big horses and let their children stroke their soft noses, then after a few minutes they would wander off to look at something else, but others were interested in what we were doing, wondering why we were using horses and why we were harvesting in the old-fashioned way. Frequently we would get people who had stooked sheaves or had led a horse from stook to stook when they were young, some having grown up in rural areas, whereas others had been evacuees from the cities during the war. Occasionally some of them fancied having a go and would fork a few sheaves up onto the load. Whereas most people tired after a minute or two, there were a delightful few who had obviously done the job before, who would fork the rest of the load, landing every sheaf the right way round and just where you wanted it. Then there were the people who had interesting stories to tell, often related to rural life, farming or horses, and others who knew that they were witnessing something special and were intrigued to know how horses work in a modern context. From all these people we gained snippets of information, another view on life, and, as they went on their way, an added impetus to keep on at what is a hard job.

Three of Geoff Morton's horses with a big load of sheaves on this 18 foot trailer.

I can only guess what most of the visitors got out of their visit, but few can have failed to realise that the horses were not some last-minute addition to provide a tourist attraction but were an integral part of the whole farming operation. This is a crucial difference, for the visitors were not seeing a staged microcosm of yesteryear but were witnessing experienced working animals doing a real job like it was no big deal. For those who work horses, the difference is more profound, because although in some respects horses are just another source of power, when you work horses all the time you initiate a series of changes which constitute a fundamental alteration in the life of a farm.

When you work horses on the land, when the horse puts its foot down on the soil, that bit of soil is squashed. In between the footmarks, the soil stays just as it was. By contrast, the wheels of a tractor leave a continuous band of compressed soil behind them and, especially when wet, smear the soil as the wheels slip, forcing the soil's open crumb structure into a paste which dries into a hard, impenetrable layer. With repeated passes of heavy machinery, it takes increasing amounts of time and power to get the land back into a suitable condition to sow seeds, and the land needs to be worked more deeply to ameliorate the compaction. If that deeper cultivation includes ploughing, the upper layers of soil with all the bacteria which thrive near the surface are buried, whilst the less biologically active soil from lower down is put on the top. With horses, soil compaction is minimal, so the cultivations only need to be shallow, requiring less power and fewer passes to make a seed bed.

European Champion horse logger, the late Mattias Rensing, driving four horses on a weeder.

Nonetheless, when working at the slower speed of a horse, the work must be spread over a longer period, so growing crops where all the work comes at once, or putting the whole farm into one crop, is unrealistic. Choosing the right mix of crops to spread the work throughout the year avoids excessively busy periods with the risk of overworking the horses and keeps them active during slacker periods. Unlike a tractor, horses need to be fed when they are not working, so the more days they are at work, the more economical they become. Whilst they are in regular work, even if working half days, they are also fit enough when busy times do come. When they are fit, resting times are shorter and less frequent which means more work gets done, and because the dirt and dead skin are loosened by sweating, active horses don't need to be groomed as much either. Horses that are worked regularly also know the routine; they go in and out of the stable on their own, they stand quietly while you harness them or hitch them up, so your job is quicker and easier too.

Once you grow a variety of crops, you end up growing them in rotation, reducing the occurrence of disease because there is less opportunity for the disease organisms to take hold from year to year. Because you need grass for the horses, you probably introduce a temporary grass and clover ley as part of the rotation. While the grass is growing, the earthworms remain undisturbed and increase in numbers, incorporating the organic matter from the manure and trampled leaves, producing a stable crumb structure in the soil. This allows water to percolate between the large pore spaces whilst retaining moisture in the smaller ones, so the land becomes less prone to drought and, in a wet time, is less likely to become waterlogged. To use the grass to best advantage, you probably add some cattle or sheep, which, when grazed with the horses, lower the number of intestinal worm eggs on the grass and the worm burden of the animals, because many of these parasites are species specific. This in turn lowers your worming costs, and with a wider range of crops and livestock you are less exposed to fluctuating prices and poor yields of one crop. By the time the grass ley is ready to be ploughed in, the foals which were born when the grass seed was sown are now working half days, and in a year or two can

An Amish man discs and rolls with six working mules in the Pennsylvania landscape.

be sold as experienced working horses or can replace an older horse. Those horses should keep working for another dozen years or more and provide you with more foals, whilst your machinery, pulled at horse speed, will last even longer. This will save you money, but not nearly as much as avoiding buying new tractors and endless tanks of fuel. Instead, the fuel for your horses is grown on the farm at the cost of growing it, and rather than the waste being toxic pollutants, it goes back to the soil to grow the next crop and feed the worms whose work will make your plough pull more easily next time.

The late Roger Clark ploughing with a Suffolk and a Percheron on his Suffolk farm.

These are just some of the many interrelated reasons why we might want to use working horses, and why we should use working horses. However, there is another reason, one which is recognised if not always understood, even by the casual observer - because they belong in the landscape. The reason it looks right and feels right is because it is right.

A landscape devoid of animals is a lonely place. We know that we can grow crops using chemical fertilisers, where for endless miles there are no animals to be seen, but there is no joy in that. So, in the spaces between fields where there are a few trees, ponds or patches of weeds, we delight in the birds and butterflies that find a home there. Once we have animals in fields between the arable crops, we have an additional focus for our attention and a landscape which is not only more interesting but is also more dynamic, because of the beetles, grubs and worms which feed on the manure, and the birds which feed on them, and the other birds which feed on them. Whilst grazing animals help bring us closer to nature, working animals bring us closer still, because we spend more time with them and develop an understanding and a relationship which is more intimate. Whilst working, we see their exertions; we notice how the ground conditions change, we see their ears and heads move as they focus on a distant or questionable object, so we learn a greater awareness of our surroundings, and because the work is done at a walking pace, we are brought back to our own biological pace.

Appreciating any environment is best done slowly because there is time to observe, time to think, time to realise that a steady, consistent approach is the appropriate way to react with nature and natural processes. It is this pace, this steady manner that is neither flashy nor glamorous which people intuitively recognise as being of deep-seated value; when they witness the draught horse leaning into the collar, they see the embodiment of this approach. Just by doing what it does, by putting one foot in front of the other, the draught horse exemplifies what is needed in rural life and rural policy - a stability based on continuity and an interest in the future, an allegiance to place and a connection to the natural world.

That stability, that connection, however, does not appear out of nothing. It grows when we put down roots, when we respect our surroundings and trust our neighbours, with the hope and expectation that they will do the same. This takes time, a steadfast conviction and an imagination that allows us to envision a future that is better than the present. If, for instance, we can imagine the trees from which our unborn grandchildren will pick fruit, if we can visualise how our local landscape would look with orchards around every corner, and if we know which plants the bees need to keep them in nectar when the trees are not in bloom, then we start to understand what stability involves, and with it, beauty and hope for the future.

The building blocks of that stability are commitment and understanding, imagination and the ability to work together, which by obvious coincidence are the very same capabilities that we need to effectively work horses.

The strong resonance which working horses have in our hearts and minds, whether we are involved with them or not, is because these same capabilities are also the most cherished part of human life. Although narrowing our focus and making quick decisions may be useful in a competitive world, it is through commitment and understanding, imagination and working together, or to say the same thing in different words, it is through involvement and kindness, creativity and community that we are of greatest value to our fellow human beings and where much of the joy of life is to be found.

Opposite page - Phil Wilson, obviously at ease and enjoying driving the author's Percheron mares, whilst Lucy, still young, is still trying to work something out!

Chapter Twelve

The Spark

Heavy Horse World, 2016

I have a neighbour who keeps cart horses, Shire horses to be precise. Now in his nineties, he was one of the last people to work horses as a young man in the 1940s. He then became a dairy farmer, but after buying a mare in the '70s he had a couple of decades of successful showing. Sometimes he talks about his early days, but he is never more alive than when talking about his experiences in the show ring with his favourite mare, who whilst not big by today's standards, was a really good type of Shire. She always went well; as he described it, when she went in the ring she would "stand another hand higher." As he related this tale, his eyes brightened, his face became animated, and not a tall man himself, he too seemed to stand half a hand higher! His enthusiasm was so infectious that it even made me smile.

I say 'even,' because I am not a great show-goer. As a friend and turnout competitor commented when I last went to the Shire Horse Show, "I thought this would be a bit too shiny for you," and he was right. I do, however, see the point of shows and am glad they are there, especially the breed shows which do a useful job in allowing you to see some of the progeny by stallions you saw a couple of years ago, and build up an overview of the direction breeding is going. Then, of course, there is the social aspect too, and for those who like competition, the prospect of winning. Still, it is not really for me, because I am not one for dressing up, glitz or competitions.

I am much more interested in the process, in the doing, and understanding how it fits in with everyday life. Although I am pleased that there are others who love to watch the horses doing their stuff, being a bystander has never really appealed to me; I have always wanted to be part of the picture, to handle the lines, to hear the jangle of the chains and the swish of the soil, to feel the sun on my back, and if needs be, the rain on my head, sometimes even right through to my skin.

This was something I knew from the first time I had any contact with working horses, at a place run by a charity devoted to preserving the draught horse. After years of wondering, I thought this introduction would either confirm my nascent interest or I could put my nagging curiosity behind me for once and for all. Within a couple of hours, I was hooked. My next experience on Geoff Morton's farm was altogether different; it was like coming home. In a way, that is just what it was, as I grew up on a Yorkshire farm, but now I could see how all this horse stuff fitted in with what I already knew. The big difference from the first place was that things happened here because they needed to happen. The fact that the horses found their own way into their standings in the morning, that all the harness was just where it needed to be behind the horses, and when they went out of the yard they just stood there waiting for you to yoke them up, all this spoke of contented routine, regular work and normality. Here, the horses were at the centre of everything, and where others may have seen a hard way of life, the horses, at least in their attitude to work and to those of us who worked with them, seemed to know different.

Whilst working on the farm, I came to realise that there are many good reasons to work horses, but for anyone who wants to do it themself, the arguments are of secondary importance; it is why you want to do it and what it feels like when you are doing it that matters. This will be different for each one of us and may lead in different directions, into the woods, growing vegetables or showing, but for me, the crux of it is doing something that needs to be done, because I can see the beauty, the practicality, the economy and the sense of it.

After years of following horses and despite being very familiar with the process, I still find working with horses to be remarkable and extra-ordinary. To be able to use a horse that is way more powerful than me is remarkable in itself, to be able to do the same work as a much more powerful tractor, though often more slowly, is a lesson in sustainable efficiency, whilst developing the close bond of collaboration with a horse is not just a joy but a profound step towards reintegration into the natural world.

From the outside, working with horses may look easy; indeed, the basics are quite simple, but like any craft requiring knowledge and skill, working with horses is a balancing act. Over time, that balance has become easier to maintain, even though I am more aware of its complexities, and it continues to demand my thought and awareness. My increasing ease is partly a product of experience, but also, having made the decision to use horses and organised everything around them, I need to use the horses to do the work. In doing that work, over days and weeks and years, as the lessons of working in harness are continually revisited and refined, it has become a contented normality for the horses and for me, both an everyday part of life and a source of deep satisfaction.

Recently, I had a visit from some people who came to see how they might use horses for growing vegetables, and for a day we tried out different pieces of equipment, they got to drive my older horse and to feel what it was like to handle the cords and steer the hoe. Whilst we unharnessed the horse at the end of the day, one of them said, “we have had a wonderful day.” To clarify that she wasn’t just being polite, she continued, “it’s like work for your soul?” The upwards inflection of the last word indicated that she wanted to know that I understood, but I knew very well what she meant.

During that single day, we had scarcely scratched the surface of horse work, of the skills of driving, of adjusting harness or assessing loads, never mind the variables of making hay or ploughing. But in those few hours she had grasped the special nature of this activity, the remarkable quiet magnificence of working in close cooperation with this exceptional species and of working with the rhythms of nature.

For the twenty-first century, our day, admittedly, was somewhat unusual, but within the realm of practical horse work, we had done nothing remarkable or out of the ordinary. We had just stood about adjusting equipment and talked about how things worked with the horses and the lines, the harness and the equipment, and looked at the soil after we had tried something different while the horses stood waiting patiently in the spring sunshine.

References

Chapter Three - Making Hay with a Single Horse

1. Estimating Draft; Steel and Rubber Tyred Wagons, by Tim Harrington, Richard Roosenberg, Dulcy Perkins and John Sarge. Rural Heritage Magazine, Holiday edition 2001.
2. Estimating Draft; Sickle Bar Mowers, by Tim Harrington, Richard Roosenberg, Dulcy Perkins and John Sarge, Rural Heritage Magazine, Spring edition 2002.
3. The Horsedrawn Mower book, by Lynn R Miller, 2007.

4 and 7. The Production of Hay on Racks and Tripods, a study by the National Institute of Agricultural Engineering, started in 1954.

5. Ron Creasey: Last of the Horselads, by William Castle, Old Pond Publications, 2012.
6. Fertility Farming, by Newman Turner, Faber 1951.
8. Haying with Horses, Lynn R Miller, 2000.
9. www.schaffmatpaerd.com/en/test-reports/

Chapter Seven - Ploughing with a Single Horse

1. A Few Long Furrows on Horsedrawn Tillage, by Eric and Anne Nordell, Small Farmer's Journal, 1993, and reprinted in 'Weed the Soil, not the Crop,' self-published by the authors.

Chapter Eight - You Can't Work Horses and Send Men to the Moon

1. https://www.woodlandtrust.org.uk/media/4808/pontbren-project-sustainable-uplands-management.pdf
2. Out of the Earth, by Louis Bromfield, Harper, 1950.
3. 'Humussphäre,' - Projekt zu energiesparenden und humusaufbauenden Methoden in der Landwirtschaft. Abschlussbericht, 2013, (Project on energy-saving and humus-building methods in agriculture) by Klaus Strüber. Conducted for the Society for Agriculture and Pedagogy e.V., Bremen, under the supervision of Prof. Dr. R Horn, Institute of Plant Nutrition and Soil Science, University of Kiel.

Chapter Nine - The Pull of the Horse-drawn Plough

1. Experimental Inquiry on Draught in Ploughing, by Philip Pusey MP., Journal of the Royal Agricultural Society of England, 1840.
2. A Cyclopedia of Agriculture, by J C Morton, 1855.

Chapter Nine - continued

3. Ploughs and Ploughing, a lecture delivered at the Royal Agricultural College, Cirencester, by James Edward Ransome, 1865.

4. Ploughing and Seeding, with Special Reference to the use of the Chilled Plough and Grain Drill, a paper read before the Airth Agricultural Science Class and reprinted in the North British Agriculturalist, by John Spier, 1892.

5. Standard Cyclopedia of Agriculture, 1909 edition.

6. Personal Correspondence with Leon Wengard of Pioneer Equipment, 2020.

7. Double Furrow Ploughs, a talk given to the Framlingham Farmers' Club, by James Edward Ransome, 1872.

8. Hornsby and Sons, sales catalogue, 1891.

9. Horsedrawn Plows and Plowing, by Lynn R Miller, 2000.

10. Pamphlet 'Horses-Mules, Power-Profit,' by the Horse and Mule Association of America, 1928.

Three of Geoff Morton's horses viewed from the seat of the hitch cart.

Recommended Reading

For anyone wanting to learn about using work horses in harness, the most comprehensive introduction is Lynn Miller's 'Workhorse Handbook.' An older book, which also contains useful information is 'The Draft Horse Primer' by Maurice Telleen. Both are by American authors so some of the practices, particularly relating to harness, differ from what is used in Britain. For a British slant, the 'Heavy Horse Manual' edited by Diana Zeuner is a good start. Following on from the Workhorse Handbook, Lynn Miller's, 'Training Workhorses; Training Teamsters,' is invaluable for any horseman or would-be horseman, particularly for understanding horse behaviour, even if you never intend to train a horse.

As an introduction into the traditional role of the farm horse in Britain, but from a modern non-farming perspective, Paul Heiney's 'Pulling Punches' tells how he started to work on Roger and Cheryl Clark's horse-powered farm in the 1980s, before he got his own small place which he wrote about in 'Farming Times.' Books about individuals who work horses, including Geoff Morton and others feature in the 'Princes of the Plough' by Barry Cockcroft, from the 1970s, while 'The Last Horseman' by Graeme Thompson details the work of John Dodd from Hexham.

First amongst the books with a knowledgeable historic perspective on working horses are those by George Ewart Evans, especially 'The Horse in the Furrow,' detailing the farming practises of East Anglia. For a similar slant on East Yorkshire is my own book 'Ron Creasey: Last of the Horselads,' which is based on Ron's memories of being a farm horseman just after the Second World War. The same geographic area is covered by Stephen Caunce in 'Amongst Farm Horses,' though his is a broader, and a more academic view, though highly readable, focusing on the whole system of hiring young men for a year at a time, including the practical and social aspects of the horselads' lives. A detailed history of the English cart horse is to be found in Keith Chiver's tome, 'The Shire Horse.' As you might expect from this old-Etonian, Cambridge-educated Church of England vicar, this is very much an establishment view, it goes into perhaps excessive detail about Shire Horse Society matters, but it is well researched and gives a detailed account of what was happening in the world of the cart horse over two centuries.

Although not the main focus of this book, the land and soil are inextricable themes when thinking about farm horses. In the early to mid-twentieth century, Sir Albert Howard and Lady Eve Balfour were in the forefront of seeing the connections between human health, food and the soil. Their observations, practical experiments and conclusions are outlined in their books, 'An Agricultural Testament' and 'The Living Soil,' respectively. Back then, they could show the effects of biological farming but could not demonstrate the mechanisms of how it happened, their work being sidelined by the powerful, profit-orientated chemical establishment. More recently, modern science has discovered the processes that underpin their observations, heralding a different approach to farming which is still in its infancy. 'What Your Food Ate,' by David Montgomery and Anne Biklé, explains how these interactions work, whilst American farmer Gabe Brown, in his book, 'Dirt to Soil' describes his journey from a chemical-dependent struggling farm to profitable Regenerative Agriculture.

Excellent and detailed research into various aspects of draught and the development of new horse-drawn machinery done by Paul Schmidt of the organisation 'Schaff mat Päerd' can be found, in English and other languages, at www.schaffmatpaerd.com, as well as in his articles in the Small Farmer's Journal, and, in German, in Starke Pferde magazine.

For those who want to work horses at all manner of farming activities, Lynn Miller has written a number of books on specific aspects of farm work. The content of 'Haying with Horses,' 'Horsedrawn Plows and Plowing,' 'Horse Drawn Cultivators,' and the mower and harness books, are obvious. I also like Stephen Leslie's 'Horse-Powered Farming for the 21st Century,' a great resource, particularly for anyone wanting to grow vegetables using horses.

A huge amount of information relevant to working horses and their role in farming and logging is to be found in the Small Farmer's Journal magazine. You can view many past articles and buy a subscription at smallfarmersjournal.com

The author mowing hay with Molly and Lucy in 2019.

Glossary

Abreast - two or more horses hitched side by side.

Ackerman steering – the type of steering where the wheels stay where they are in relation to the vehicle, like on a car, rather than a whole axle pivoting in the centre.

Amish - a Christian group mostly found in North America. Members of the Old Order Amish churches are mostly involved in farming, relying on horses and mules for work and transport.

Angle of draught – the angle of the traces and/or draught chain between an implement or vehicle and the horses' hame hook.

Ardennes, or Ardennais - a breed of compact draught horse originating in the Ardennes region of France and Belgium.

ATV - all terrain vehicle, quad bike.

Back band – the strip of leather or webbing that goes over the horse's back, where a saddle might sit, that holds up the traces.

Beam, of plough – the longitudinal piece of wood, iron or steel, to which the parts of the plough are attached.

Belgian, American Belgian – a breed of horse originating in Belgium which were exported to America in the late nineteenth and early twentieth centuries. Over time, a greater proportion of American Belgians have been bred to have the sorrel colouring, and since the 1950s they have typically increased in height and have lost bone and substance.

Binder – the reaper-binder, a machine that cuts the corn and ties it into sheaves.

Biodynamic – a type of Organic farming which emphasises the use of planting and harvesting dates in relation to celestial bodies and the use of preparations to enhance the function of the soil.

Bodkin fashion – a method of yoking three horses, where two are in line, one in front of the other, usually walking in a furrow, with the third on the land. In this case the point of attachment to the plough is one third of the way along the evener, the two furrow horses pulling on the same swingletree.

Body – the part of the plough, usually of cast metal, to which the main working parts, the share, land-side and mouldboard are attached.

Brabant – breed of heavy draught horse, also known as the Belgian. In America especially the word Brabant is used for recent imports of the Belgian breed as it is bred in its country of origin, often roan in colour and heavily built.

Breast – the mouldboard of a plough.

Bridle – the harness which sits on a horse's head, to which the bit is attached.

Britchen (breeching) - the part of the harness that goes round the horse's back end, including the hip and loin straps that keep it in place. The britchen allows a vehicle to be slowed down, stopped and reversed.

Buck back rope – used on a strung-out team, this is the rope (or leather strap) which connects a wheeler's bit to the draught chain which comes from the leaders, causing him to slowdown or stop as the leaders slacken their pace or stop.

Buckrake – an implement with forward facing tines which run along the ground to gather hay.

Cart – a two-wheeled horse-drawn vehicle.

Carting – to move stuff in a cart, but also used generically to move stuff about.

Cart harness - harness used with a horse in shafts, including a cart saddle and britchen.

Cart horse – during the heyday of the horse, this was the most common term in Britain for a heavy draught horse, also known as a plough horse, farm horse or dray horse.

Centre of draught – the point marking the centre of resistance on an implement.

Cereal – a grain crop, typically of oats, rye, wheat or barley.

Chains, chain harness – the harness used in Britain for ploughing and other work where a britchen is not required.

Chain harrows – flexible harrows consisting of an interlinked mesh of steel, each section ending with a downward facing end which forms the tine.

Clydesdale - draught horse breed native to Scotland, but which was also common in Northern England and Ireland. They exhibit many of the same characteristics as the Shire, sharing much common ancestry from the the time before the breeds were established. The Clydesdale was traditionally more lightly built and had a faster walk than the Shire. Since the mid-twentieth century, they have become taller and lighter.

Cock – a small pile of hay made in a field to minimise the hay's exposure to the weather as it dries.

Cobbletree - (Yorkshire) the wooden or metal evener which is used to equalise the pull from two horses. In different parts of Britain also known as a pommeltree, whippletree or master tree, in America as a doubletree.

Collar – the leather and straw ring which sits on the horse's shoulders into which he pushes, transferring his efforts to the hames and traces and thence to the implement.

Collar pad – a pad which sits under the collar, usually made of deer hair covered in cloth, but sometimes of synthetic materials. Used with American collars.

Cords – hemp ropes which are used as reins to steer horses.

Corn - in Britain, and in this book, this refers to oats, wheat, rye and barley.

Coulter - the part of a plough which cuts the vertical cut in the soil.

Coulter - the part of a seed drill which opens up a groove into which the seed is dropped.

Coupling band – a short length of thin rope which connects horses together, from one bit ring to the neighbour's bit ring.

Cover board - on American ploughs, this is like a small mouldboard which sits above the mouldboard proper. Because it has a sharper curve it puts the soil and vegetation from the top corner of a furrow slice into the bottom of the furrow, in the same manner as a skim coulter.

Crown – the first part of a ploughed land when the first two big furrows have been turned together, also known as a rigg, ridge or top.

Crumb structure – the nature of the soil where small particles are held together in relatively stable bigger particles.

Crumbler – implement like a roller but with gaps between each ring, or like a mesh, used to refine and consolidate a seed bed.

Culti-mulcher – implement which combines cultivating tines and rollers or crumblers to create a seedbed.

Digger – a plough type with a sharp-turning short mouldboard which crumbles the soil as it turns.

Disc coulter – a revolving disc of steel which cuts the vertical cut when ploughing.

Dolly wheel – a wheel which holds up the rear of a pole on an implement, minimising the weight at the front of the pole which otherwise is transferred to the horses' collars and necks.

Draught – the force needed to move a load.

Draught bracket – the place on an implement to which the draught chain or evener is attached.

Draught chain – a chain which connects the horses' eveners to the implement.

Draught horse – a type of horse used for pulling things. Nowadays, and in this book, this term is used to mean a heavy draught horse.

Eveners – the pieces of metal or wood which keep an individual horse's traces away from its sides, and those which equalise the pull from the individual horses in a team.

Fingers – the pointed projections along a mower's cutter bar, which hold the grass in place whilst the knife sections move back and forth providing a scissor action against the fingers.

Fold yard – an open or roofed yard where animals are kept, usually during winter.

Forecart - hitch cart.

Frog, of plough – the front part of a plough body onto which the share is attached.

Furrow – a ribbon of soil turned by a plough.

Furrow - the trench along the edge of ploughed land, where a strip of soil has been turned to the side leaving an open furrow.

Furrow horse – the horse which walks in the furrow whilst ploughing.

Furrow slice – the strip of soil turned by the plough.

Furrow wall – the vertical side of the open furrow.

Furrow wheel – the wheel of the plough which runs in the furrow.

Gang plough - a ride-on plough with two or more plough bodies.

General-purpose plough – a plough with a relatively slow-turning mouldboard.

Glume – the thin outer casing covering a grain of corn.

Grass board – the board at the end of a mower's cutter bar which turns the far end of the swath back towards the mower, leaving a clear piece of ground next to the standing grass.

Hake – (Am. clevis) the crescent shaped attachment at the front of a plough to which the draught chain is attached.

Halter – piece of harness which goes on a horse's head, used to tie it up and to lead it.

Halter shank – lead rope, rope which attaches to the halter to tie or lead a horse. Also used in Yorkshire when hitching two or more horses abreast to regulate their speed and the distance between the horses.

Hame – the wood and/or metal part of the harness with sits in the groove of a collar and transfers the pull, via the traces to the implement.

Hame ring – a ring on the hame, through which the reins or cords are passed to keep them out of the way of a horse's legs and clear from the rest of the harness.

Hay grab, hay grapple – a metal grab used to raise loose hay onto a stack with a rope.

Headland – the edges of a field where the horses or tractor turn. When ploughing, the headland is left unploughed until the rest of the field is done.

High cut work – also known as oat-seed furrow – a type of ploughing where the top surface is left in steep-sided ridges. Historically used when broadcasting oats, now only used in competition ploughing.

Hitch – the way horses are fastened together in a team, or a number of horses working together.

Hitch cart – two wheeled vehicle, usually, which allows horses to be used with implements having a hitch, rather than each implement needing a pole or shafts.

Hock – the joint in a horse's back leg which equates to our heel.

Horseman – in Britain, the usual name for someone skilled in the use of horses. In America, the non-gender specific term 'teamster' is used. In my use of the word horseman I include anyone practised in the use of horses.

In line – a way horses are hitched together when one is in front of another.

Knife coulter – the knife-like part of a plough that makes the vertical cut which will form the furrow wall.

Land– a division of a field when ploughing.

Landau - a four-wheeled carriage with a two-piece top which can be folded away, the driver sitting on a raised seat outside.

Land horse – when ploughing, the horse which walks on the unploughed land.

Landside – the part of the plough that runs up against the furrow wall.

Land wheel – the plough wheel that runs on the unploughed land.

Leader – In a team of horses, where there are two or more rows of horses, the leaders are the ones at the front.

Leading – (colloquial Yorkshire) moving stuff by vehicle or sledge, the same meaning as carting.

Lead chain – the long chain from the leaders of a multiple team, which goes back to the vehicle or to the rear set of eveners.

Legume – a plant of the Leguminosae family, such as beans, peas, vetches or clover, typified by their root nodules which contain the bacteria which fix nitrogen from the air and turn it into a form that the plants can use.

Ley – a temporary crop of grass and/or clover grown in between a sequence of arable crops.

Line – the length of leather, synthetic material, or rope which goes from the horses' bits to the driver. In Britain we also use the word rein.

Loam – a soil which contains a mixture of clay and sand particles.

Loose hay – stored hay which is not baled.

Manger – trough for feeding animals.

Mangold – large swollen root crop, part of the beet family, used to feed livestock.

Mouldboard – part of a plough that turns the furrow slice.

Nearside - the horse which is on the left when viewed from the driver's position behind the horses.

Neck yoke – the piece of wood or metal fixed at its centre to the front of a pole, the ends being attached to the horses' pole straps, allowing them to steer the pole and to hold back the vehicle.

Nicker – a soft whinnying sound.

No-till – the process of farming arable crops where the land is not cultivated.

Offside - the horse which is on the right when viewed from the driver's position behind the horses.

Open bridle – a bridle which does not have blinkers.

Organic – a method of farming which uses natural processes to provide the nutrients for crops and animals, rather than manufactured soluble fertilisers.

Organic matter – the part of the soil which is made from decaying plant and animal matter, containing carbon.

Pawl – a spring-loaded piece of metal which engages in a cog, so the cog can push the pawls and transfer the rotating force when moving in one direction, but when the cog turns in the other direction, the pawls spring out of the way.

Packer – like a roller, but with wider spaces in between the ridges, used to consolidate cultivated ground.

Percheron – breed of draught horse originating in the old French province of Le Perche. Many were imported into north America in the late nineteenth century, becoming the most common draught horse in the US by the 1930s. The American Percheron today tends to be taller and lighter than the French horse, with an exaggerated trot. The Percheron has been bred in Britain since 1916 and is less extreme than the American type, though there have been some imports from America recently.

Point – the share of a plough.

Plow (Am) - plough.

Pole – the piece of wood or metal attached to the front of a vehicle or implement which passes between two horses to a neck yoke, allowing the vehicle to be steered and stopped.

Pole strap – the part of harness which attaches the neck yoke to the horses' hames or collar.

Power cart - hitch cart with a mounted engine.

PTO shaft - power take-off shaft, the rotating shaft which transfers power to an implement from the back end of a tractor, or from a hitch cart with ground drive or a mounted engine.

Reed comber – an attachment to a threshing machine which combs the straw.

Rulley, rully – a four-wheeled vehicle without sides.

Regenerative Farming– a method of farming which aims to regenerate the organic matter levels, the structure, the biology and the health of the soil.

Rein - ribbon of leather used to steer a horse.

Reversible plough – a plough with both left and right-hand bodies, used alternately so all the land is turned in one direction, no matter which direction the plough is travelling.

Rigg (northern) – ridge, both a geographical feature and the first part of ploughing where the first two furrows are ploughed towards each other.

Rotation – a sequence of crops grown to balance the nutrient requirements of each crop and to minimise the build-up of disease.

Semi-digger – a plough where the mouldboard is steeper than a general-purpose mouldboard, but less than the severe digging bodies.

Shafts – the pieces of wood or metal that go on either side of a horse's body to hold up a vehicle and transfer the horse's efforts to the vehicle.

Share – the part of the plough that cuts the bottom of a furrow and starts to lift it.

Sheaf – a bundle of corn tied with twine, made by a binder. (Am. bundle)

Shin – the relatively sharp vertical part of a digger plough which is attached at the leading edge of the mouldboard, and which cuts the vertical part of the furrow slice, used instead of a coulter.

Shire – breed of heavy draught horse originating in England, with early influx of blood from horses from the Low Countries. The Shire is the heaviest of the British breeds and slower than the Clydesdale with more feather on the legs. Over the last fifty years they have become taller but also lighter in build.

Shoulder chain - the short chain that connects the shaft to the hame hook on British cart harness.

Sickle bar – the cutter bar of a mower.

Side-delivery rake – implement which uses rotating raker bars to push hay sideways, so turning the row of hay upside down.

Side draught – the forces on an implement, particularly a plough, which are not in the direction of travel, forces which add to the draught by increasing friction.

Skim coulter, skimmer – a small plough-shaped attachment which turns a mini furrow in front of the plough body.

Slade – the long piece of steel on a plough that runs in the corner of the furrow, resisting the force of the soil on the mouldboard and contributing to keeping the plough running straight.

Spider-wheel rake – a hay rake with a series of wheels, the extended spokes of which form tines. As the rake moves forwards, the wheels turn, and since they are at an angle to the direction of travel, they push the hay to the side, turning it over.

Stall – division of a stable for one horse, or occasionally two.

Steerage hoe – hoe mounted on wheels for weeding crops grown in rows, where the hoe is steered by the driver's hands or feet.

Stook (Am. shock) – a number of sheaves stood up in pairs to keep the grain away from the ground and to allow the corn to dry.

Suffolk, Suffolk Punch - a draught horse breed originating in East Anglia, shorter than the Shire or Clydesdale. Numbers are low. They are also bred in America though the Suffolk Horse Society has yet to accept American-bred horses in the British stud book.

Sulky plough – a single-furrow riding plough.

Swath – the row of grass left behind a mower.

Swingletree (Am. Singletree) – the piece of wood or metal which keeps a horse's traces from rubbing its sides, and which transfers the pull from the horse's traces to the implement or eveners.

Swing plough – a plough without wheels.

Team – two or more horses hitched together. (In American usage, team refers to a pair of horses. What the British would call a team, the Americans call a hitch.)

Teamster (Am.) – someone who drives horses, a horseman.

Tedder - implement that throws hay upwards and backwards to fluff it up and allow the air to it.

Threshing – separating the grain from the straw.

Tine - pointed piece of metal which moves hay, or which, when dragged through soil, separates the soil.

Tongue (Am.) – the pole at the front of an implement or vehicle which passes between the horses allowing it to be steered or stopped.

Tongue truck (Am.) – pair of wheels which support the rear end of a tongue or pole.

Topdressing – applying fertiliser on top of the soil.

Topless guards – triangular parts of a mower's cutter bar which form the shear surface against which the sections of knife move to cut the grass, used instead of fingers.

Trace - length of chain or leather which attach to the hame at one end and the swingletree at the other.

Tripod - three-legged construction to keep hay off the ground and allow it to dry.

Trip rake - implement with semicircular tines which gather up hay as it is drawn forward. When full, it is lifted, or tripped, by pressing a foot pedal which causes the tines to be raised by the movement of the wheels.

Waggon, wagon – four-wheeled vehicle, usually with sides.

Western Breeching - style of harness from America, now in common use in Europe, where each end of the breeching is joined to a quarter strap, both of which clip into the breast strap under the horse's belly. The breast strap runs between the horse's front legs and clips onto the neck yoke, so the braking force is transferred in a fairly straight line from the britchen to the neck yoke.

Wheat reed - the type of wheat straw which comes from a reed comber used for the West Country style of thatching.

Wheeler - when more than one row of horses is hitched together, the wheelers are the horses in the back row, nearest the implement or vehicle.

Wing – the part of the plough share which projects towards the open furrow

Yoke - originally referring to an ox yoke, but used as a verb meaning to hitch up.

About the Author

William Castle is a violin maker, horseman and smallholder. He grew up on a farm in North Yorkshire and has always been interested in older methods of doing things, not just as an historical interest but also in relation to contemporary choices about ways of living, economy and the environment. Having learnt to work horses on a horse-powered farm, he uses his two Percheron mares to work his own smallholding.

He was a regular contributor to the Heavy Horse World magazine and still writes for the Small Farmer's Journal.

'Working horses: Tradition, Innovation and Sustainability' is his second book. His first book, 'Ron Creasey; Last of the Horselads,' was published by Old Pond Publications in 2010, and he is currently working on another book based around the seasonal work with his own horses.

More of William's writing can be found on his website (working-horses.com) where you can also buy copies of this book.

www.ingramcontent.com/pod-product-compliance
Lightning Source LLC
LaVergne TN
LVHW070952210926
848482LV00002B/191